INTRODUCTION TO WIRELESS LOCALIZATION

INTRODUCTION TO WIRELESS LOCALIZATION

WITH iPHONE SDK EXAMPLES

Eddie C.L. Chan

The Hong Kong University of Science and Technology, Hong Kong

George Baciu

The Hong Kong Polytechnic University, Hong Kong

Library of Congress Cataloging-in-Publication Data

Chan, Eddie C. L.
    Introduction to wireless localization : with iPhone SDK examples / Eddie C.L. Chan, George Baciu.
        p. cm.
    Includes bibliographical references and index.
    ISBN 978-1-118-29851-0 (cloth)
    1. Location-based services.   2. iPhone (Smartphone)–Programming.   I. Baciu, George, 1961-   II. Title.
    TK5105.65.C48 2012
    005.3–dc23

                                                                                    2011052849

Set in 10/11.5pt NimbusRom9L by Thomson Digital, Noida, India.
Printed and bound in Singapore by Markono Print Media Pte Ltd.

Contents

Preface

W ireless localization is a fascinating field at the intersection of wireless communication, signal processing, physics, mathematics and human behavior, which strives to harness different mobile platforms to navigate and broaden our horizon. This book is exclusively dedicated to the positioning system, which is the killer application of the 21st century, riding on the success of Global Positioning Systems and mobile technologies. Many commercial and government organizations as well as university campuses have deployed wireless broadband such as IEEE 802.11b. This has fostered a growing interest in location-based services and applications. This book provides a comprehensive overview of the entire landscape in both outdoor and indoor wireless positioning systems with practical iPhone application examples.

Who This Book Is For

The text you hold in your hands has a different flavor from most of the other currently available books on wireless positioning. First and foremost, this book should be readable by anyone who is in or beyond their second year in a computer science program. We assume a bare minimum of mathematical sophistication and C programming skill (or some familiarity with object-oriented programming skill).

This book is primarily intended for anyone who wants to study wireless localization. It offers an insight into the maze of mobile, positioning and AI technologies. This book covers basic formulae, algorithms and mathematical calculations involved in location-aware applications. It has been planned in a manner to benefit all those developing positioning systems, such as professionals, engineers and researchers.

This book is intended for students taking a course in wireless localization that emphasizes positioning technologies and mobile application developments. The content can be treated as the material in a course structure with many explanations of fundamental positioning techniques throughout the text, and we have added many programming examples. All programming examples are updated and developed on the iOS 5.0 platform. These are an independent, practical, fun way of learning the material presented and getting a real feel for the subject.

Features

Each chapter in this book is almost self-contained. We do not demand that the reader come armed with a thorough understanding of positioning technologies. This book guides you through understanding the signal propagation, positioning and interference concepts and algorithms. We have incorporated iPhone programming examples that help readers to understand the concepts, theories and algorithms. Readers will come away from this book with an ability to develop and implement real location-aware applications.

Themes featured in *Introduction to Wireless Localization* include:

- An accessible introduction to positioning technologies such as Global Positioning System and Location Fingerprinting

- A thorough grounding in signal propagation, line-of-sight and interference effects to the positioning accuracy

- Hands-on skill to iPhone programming for location-aware application

- An in-depth solution to some open problems in wireless positioning

Organization

There are three main parts in this book. Part I covers the Wi-Fi positioning systems (Chapter 2 to 6); Part II covers the outdoor positioning systems (Chapter 7 to 10) and finally Part III introduces the applications of wireless localization (Chapter 11 to 12). Two appendix chapters are included at the end of the book for those not familiar with the iOS Software Development Kit (SDK) and Objective-C programming environments.

- Chapter 1 – Introduction to Wireless Localization
  This chapter is an introduction and overview of the material.

- Chapter 2 – Installation of Wi-Fi Infrastructure
  Localization systems for indoor areas that make use of existing wireless local area network (WLAN) infrastructure and location fingerprinting approach have been suggested recently. Chapter 2 covers how to set up a Wi-Fi infrastructure specifically for the positioning system. It gives an overview of the pre-installation criteria, standard of the Wi-Fi positioning infrastructure.

- Chapter 3 – Algorithms in the Wi-Fi Positioning System
  This chapter covers the positioning algorithms of location fingerprinting and propagation based methods. It also includes the evaluation methods and comparisons of each WLAN positioning system.

- Chapter 4 – Implementation of Wi-Fi Positioning in iPhone
  The chapter introduces how to build a customized Wi-Fi positioning system. It includes the implementation of the algorithms in iPhone.

- Chapter 5 – Positioning across Different Platforms
  This chapter presents the signal variance issue of different mobile platform. It also solves the problem due to the signal variance from different platforms.

- Chapter 6 – Wi-Fi Signal Modeling
  This chapter introduces Wi-Fi signal modeling methods that visualize and analyze the intensity of the Wi-Fi zone for post-installation.

- Chapter 7 – Introduction of Global Positioning System
  This chapter describes the history, algorithms and components of GPS.

- Chapter 8 – Study of GPS Signal and Algorithms
  This chapter provides an in-depth study on GPS signal and algorithms.

- Chapter 9 – Differential GPS and Assisted GPS
  This chapter presents the methodologies of AGPS using the GSM and 3G cell phone networks. It also covers algorithms used in DGPS and includes the implementation of GPS system in iPhone.

- Chapter 10 – Other Existing Positioning Technologies
  This chapter introduces other existing positioning technologies such as acoustic-based, vision-based and RFID-based.

- Chapter 11 – AI for Location-aware application
  Location-aware application is not only to solve the problem 'Where am I?' but also solve more complex questions such as 'Any good burger joints around here?'. It should assist the user to make the best choice using artificial intelligence (AI). This chapter includes the implementation of an iPhone application which finds the favorable dinning place according to user preference.

- Chapter 12 – Beyond Positioning: Video Streaming and Conferencing
  While reaching the meeting place, users may need to download some location-aware information or communicating with other persons. In Chapter 12, we look at data streaming technology and communication applications in iPhone. It includes the implementation of multimedia data transmission, such as data streaming and video conferencing.

- Appendix A – Starting the iOS SDK
  This is the first appendix chapter that helps readers to get ready for the iOS SDK environment.

- Appendix B – Introduction of Objective-C Programming in iPhone
  This is the second appendix chapter that introduces basic programming techniques in the Objective-C language. Objective-C is slightly different from C language in syntax, pre-defined methods and naming of files. After you read through this chapter, you will have become familiar with the Objective-C language.

Supplemental Material and Technology Support

This book includes lecture outlines in PowerPoint for the text and program codes which are free to adopters. Periodic updates and slides to this book can also be downloaded in the web page below at

www.wiley.com/go/chan/wireless

The reader is encouraged to send any corrections to

csclchan@gmail.com

Acknowledgements

I (Eddie C. L. Chan) am deeply indebted to Professor George Baciu for his support in writing this book. He is my mentor and my friend who always encourages me. I had my knowledge enriched and gained greatly from his clarity of vision and his view of computer science.

I am obliged to Professor Mordecai J. Golin and Professor Brian Mak for providing me with a warm and friendly working environment in the Department of Computer Science and Engineering at The Hong Kong University of Science and Technology.

I obtained my Ph.D. in Computer Science at The Hong Kong Polytechnic University (PolyU). Studying at the PolyU was not only a turning point in my life, but also a wonderful experience.

I would like to express my gratitude to my friend, Mr. Tony Ao Ieong for providing me with a wealth of ideas about positioning techniques. He is an integral part of this book. Mr. Raymond Ho Man Chun gave me advice to design the layout of book. I would like to thank him for the excellent design skills he has applied to this book. I am also pleased to acknowledge the cooperation and technical support of Mr. Mak Sin Cheung.

My mother and father certainly merit mention for not only offering support when I quit my job in order to work on this book, but also throughout my entire life. They have worked very hard to raise me and give me love and support. They have shaped and molded me into the person that I am today, and my only wish is to see them happy and healthy.

I am also grateful for those whom I hold dear to my heart for the long hours that they sat and listened to the issues I had surrounding this book. I want to thank my uncle, Dr. Alfert Tsang, and his wife, Mrs. Michelle Tsang, are always good listeners and advisers. Special thanks goes to my girlfriend, Miss Kami Hui, for her encouragement and support while writing this book. I want to thank Mr. Martin Kyle for his patience, understanding, and valuable assistance with many aspects of this book, including proofreading and indexing. Lastly, special thanks must go to the publishing team, Mr. James Murphy and Miss Shelley Chow from John Wiley & Sons, Mr. Prakash Naorem and Mr. Martin Noble, for their help and support in the production process of this book.

About the Authors

Eddie C. L. Chan received his BSc, MSc and PhD degrees, all in computer science, from The Hong Kong Polytechnic University (PolyU) in 2005, 2007, and 2010, respectively. During his postgraduate study, he was the recipient of the Best Student Paper Award in the International Conference on Fuzzy Computation, Madeira, Portugal in 2009. He received the Best Presentation Award of Research Project and Alan Turing scholarship from PolyU in 2007 and 2008. He was awarded the 2nd-Class Group Award from the 9th Philip Challenge Cup in China in 2005. His work in wireless localization has been published in around 30 refereed papers. He has also participated in academic conference events. He was the local chair in IEEE WiMob 2011, session co-chair in IEEE CMC 2010 and publicity chair in IEEE ICCI 2009. He was a system consultant in Itapoia Group Limited (2007) and PTec Limited (2005). His research interests include wireless localization, communication, fuzzy logic, 3D visualization of tracking system, agent technology and data mining.

George Baciu holds a PhD and a MSc degree in Systems Engineering and a B.Math degree in Computer Science and Applied Mathematics from the University of Waterloo. He has been a member of the Computer Graphics Laboratory and the Pattern Analysis and Machine Intelligence Laboratory at the University of Waterloo and subsequently Director of the Graphics and Music Experimentation Laboratory at The Hong Kong University of Science and Technology in Hong Kong. Currently, Prof. Baciu is a full Professor and Associate Head in the Department of Computing, and the founding director of the Graphics and Multimedia Applications (GAMA) Laboratory at The Hong Kong Polytechnic University. His research interests are primarily in mobile augmented reality systems, user interfaces, physically-based illumination, rendering, image processing, motion tracking and synthesis for both outdoor and indoor location aware systems.

Chapter 1

Introduction to Wireless Localization

Don't let the noise of others' opinions drown out your own inner voice. And most important, have the courage to follow your heart and intuition. They somehow already know what you truly want to become. Everything else is secondary.

Steve Jobs
2005

Chapter Contents

▶ Introducing the background of positioning technologies

▶ Defining the open problems for positioning technologies

▶ Understanding factors leading to a successful wireless positioning system

Wireless localization in this book means 'determining the position of a user/object by wireless signal.' Determining the position of a user requires tracking techniques from indoors to outdoors. Global Positioning System (GPS) is a fully functional Global Navigation Satellite System (GNSS) developed by the United States Department of Defense. In the early days it was used as a tool for map-making, land surveying, and scientific uses. Nowadays it is more widely used. Some individuals may own a pocket PC or palm phone with GPS functions. However, GPS is limited in that it requires dedicated hardware. It is also very expensive in terms of labor, spectrum and capital costs to implement a specialized infrastructure in indoor areas solely for position location. GPS and other positioning approaches like acoustic-based and light-based are most effective in relatively open and flat outdoor environments but are much less effective in non-line-of-sight (NLOS) environments such as hilly, mountainous, or built-up areas. These localization applications have two particular disadvantages: first, they require the source to have a high intensity and

Introduction to Wireless Localization: With iPhone SDK Examples, First Edition. Eddie C.L. Chan and George Baciu.
© 2012 John Wiley & Sons Singapore Pte. Ltd. Published 2012 by John Wiley & Sons Singapore Pte. Ltd.

to be continuously propagated and, second, they can localize only within the area covered by the sound, light or FM wave signals.

Cellular Positioning System (CPS) makes use of radio waves broadcast in the cell phone towers to determine the position. It works in outdoors and indoors environments, but it can only get a rough positioning. Therefore, it usually works together with GPS. Assisted GPS (AGPS) uses the cellular network tower which installed the GPS receiver to assist users to get the satellite information.

Recently, many public places and campuses have deployed the wireless local area networks (WLANs) such as Wi-Fi – IEEE 802.11b. Wi-Fi is a wireless technology brand owned by the Wi-Fi Alliance intended to improve the interoperability of wireless local area network products based on the IEEE 802.11 standards. Wi-Fi networks have become widely deployed and are fueling a wide range of location-aware computing applications. Accurate user location information enables a wide range of location-dependent applications. A software-only solution can be integrated as a location-sensing module of a larger context-aware application on infrastructure wireless LANs. Wi-Fi positioning may be a useful supplement to any of these localization approaches.

The most widely used techniques that Wi-Fi positioning uses to locate Wi-Fi enabled devices are the propagation-based and location-fingerprinting-based (LF-based). Propagation-based techniques measure a Wi-Fi transmission's received signal strength (RSS), angle of arrival (AOA), or time difference of arrival (TDOA). Propagation-based techniques use mathematical geometry models to determine the location of the device. LF-based techniques locate devices by accessing a database containing the fingerprint (i.e., the RSSs and coordinates) of other Wi-Fi devices within the Wi-Fi footprint. These devices then calculate their own coordinates by comparing with those contained in the relevant LF database.

Figure 1.1 summarizes the equipment, working environment, power usage and resolution of these three positioning systems (GPS, CPS and WPS).

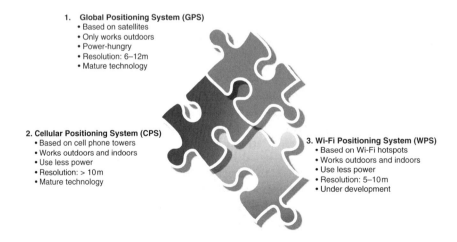

1. **Global Positioning System (GPS)**
 • Based on satellites
 • Only works outdoors
 • Power-hungry
 • Resolution: 6–12m
 • Mature technology

2. **Cellular Positioning System (CPS)**
 • Based on cell phone towers
 • Works outdoors and indoors
 • Use less power
 • Resolution: > 10 m
 • Mature technology

3. **Wi-Fi Positioning System (WPS)**
 • Based on Wi-Fi hotspots
 • Works outdoors and indoors
 • Use less power
 • Resolution: 5–10 m
 • Under development

Figure 1.1   Overview of positioning systems.

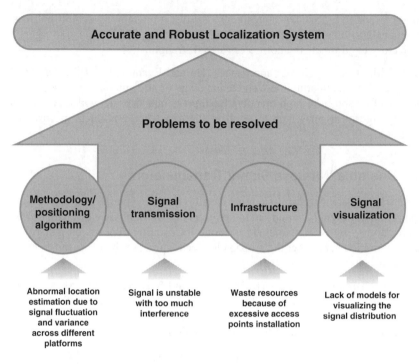

Figure 1.2   Four major problems of the localization system.

1.1   Open Problems in Positioning Technologies

Although there has been a large amount of research on positioning systems, the accuracy and robustness of these systems are still not entirely satisfactory and many open problems have not yet been solved. There are four major problems: (1) computational-intensive and inaccurate positioning algorithms, (2) unstable wireless signal transmission, (3) interference caused by unstructured WLAN infrastructure and (4) lack of models for visualizing the signal distribution. Figure 1.2 depicts the four major localization problems.

1.1.1   Inaccurate Positioning Algorithms

Existing Wi-Fi positioning algorithms (Kaemarungsi and Krishnamurthy, 2004; Kwon et al., 2004; Taheri et al., 2004; Li et al., 2005) suffer from the high computational complexity of location estimation. These algorithms perform signal sampling, weighting, and filtering. Some algorithms (Tiemann et al., 2000; Chen et al., 2004; Wang et al., 2006) make assumptions, such as assuming that WLAN signals obey a Gaussian distribution, even while it has been shown that WLAN signals very often obey a left-skewed distribution (Bahl and Padmanabhan, 2000). Another unsatisfactory assumption is that no loss of energy arises from signals being reflected by obstacles.

Common positioning approaches such as GPS (Taheri et al., 2004), acoustic-based (Chen et al., 2004; Stoleru et al., 2005, 2006), and light-based approaches are the most effective in relatively open and flat outdoor environments but are much less effective in non-line-of-sight (NLOS) environments such as hilly, mountainous, or built-up areas. Some filter signal algorithms (Liu and Chen, 2004; Chang *et al.*, 2007) use a convergence factor of a trajectory to eliminate the noise from signal strength.

Sequential Monte Carlo (SMC) approaches (Hu and Evans, 2004; Dil *et al.*, 2006) estimate locations by calculating the weighted average of all signal samples. However, these are not effective and have high computational costs in sensor networks. An optimization or filtering algorithms may help to improve the dependency problem of wireless signal transmission.

1.1.2 Unstable Wireless Signal Transmission

Unstable wireless signal transmissions are usually caused by interference and by wave propagation problems. Signal overlaps and interferences can seriously worsen the performance of localization systems. A stable and accurate localization depends on a stable wireless signal transmission yet neither the propagation techniques nor the LF-based techniques can guarantee that.

LF-based localization techniques (Kaemarungsi and Krishnamurthy, 2004; Taheri *et al.*, 2004; Li *et al.*, 2005; Fang *et al.*, 2008; Kjaergaard and Munk, 2008; Swangmuang and Krishnamurthy, 2008) require an initial survey with a very large training dataset and each signal sampling is very sensitive to signal fluctuation.

Propagation-based localization techniques (Prasithsangaree *et al.*, 2002; Jan and Lee, 2003; Kwon *et al.*, 2004) must compute every condition that can cause a wave signal to blend.

Both techniques are strongly dependent on stable wireless signal transmissions. At the same time, the common unsystematic approach is to place more APs to improve coverage. However, this may still leave blind spots where there are too many access points packed too closely together. This can lead to signal fluctuation and overlap, which is both wasteful and can cause interference (Budianu *et al.*, 2006)

1.1.3 Unstructured WLAN Infrastructure

An unstructured approach to WLAN infrastructure design implies poor resource utilization (Budianu *et al.*, 2006) WLANs are typically made up of many access points (APs) or nodes. These access points (APs) are manually placed and positioned on the basis of measurements of RSS (received signal strength) taken by engineers empirically. For example, if wireless access points are placed too close together, their signals can overlap.

Not only can this cause interference, but it also poses a potential security risk and it increases the cost of the installation as in certain cases fewer access points could be used to achieve optimal coverage. If, on the other hand, the access points are placed too far apart, there will be a potential increase in the number or extent of flat spots or weak signal areas, which can cause connections to become unusable.

WLAN infrastructures are installed and operated in two configurations. The first configuration is in the absence of wireless access points, communicating directly with each other in a peer-to-peer style. The second configuration makes use of wireless access points where all devices on the wireless network communicate with each other and services provided by the network operate through these access points. This configuration is dominant and used by all network providers in a structured environment. However, peer-to-peer connections constantly change as nodes are removed and thus coverage cannot be guaranteed.

To provide wireless coverage in a particular area, several wireless access points are placed in strategic positions and emit a signal that the clients use to communicate. One of the main issues concerning this second configuration is the placement of the base stations so as to ensure that optimal coverage is provided, meaning that the number or extent of 'flat spots' (where no signal is present) is minimal.

1.1.4 Lack of Signal Analytical Models

The task of localization is not limited to location estimation but is also carried out using analytical models. For example, WLAN signal analytical models can be used to visualize the distribution of signals and help to improve the design of positioning systems by eliminating WLAN access points (APs), shortening the sampling time of WLAN received signal strength (RSS) in location estimation, and ensuring that all vital areas of a building have wireless coverage.

Currently there are very few support tools available for planning the installation of APs, or to visualize and monitor signal coverage of the installed network. The available tools are designed only for outside installations, where the wireless signal strength and Global Positioning Service information can be combined to provide feedback.

There is a lack of visualization models that can be used as a framework for designing and deploying positioning systems. These models should have spatial elements for visualizing the RSS distribution, evaluating and predicting the precise performance of indoor positioning systems based on location fingerprinting. Such an analytical model would support the placement of Wi-Fi APs so as to achieve the maximum throughput.

Most existing methods for modeling location fingerprinting (Kaemarungsi and Krishnamurthy, 2004; Swangmuang and Krishnamurthy, 2008) depend on the accurate performance of positioning systems and proximity graphs, such as Voronoi diagrams and clustering graphs. They usually make use of the Euclidean distance to determine positions. Some researchers (Fang *et al.*, 2008; Kjaergaard and Munk, 2008; Swangmuang and Krishnamurthy, 2008) have ignored radio signal properties and others have assumed the distribution of the RSS is Gaussian and pair wise. Yet Bahl and Padmanabhan (2000) have shown that the distribution of the RSS is not usually Gaussian but rather left-skewed and the standard deviation varies according to the signal level.

1.2 Factors Leading to Effective Positioning Systems

There are five major factors (Figure 1.3) that must be considered in the design of an efficient and effective localization system: (1) an accurate positioning algorithm, (2) a stable WLAN signal transmission, (3) a structural WLAN infrastructure that could support intensive tracking, (4) a model that can visualize the WLAN signal distribution to prevent signal black spots and interference and, finally (5) a retrieval system that can provide location-aware information that matches or responds to user needs. The following briefly describes each of these factors in greater detail.

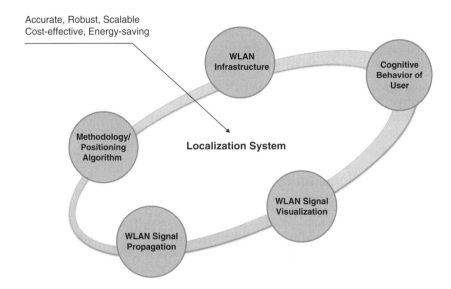

Figure 1.3   Five major factors of WLAN localization system.

1.2.1   An Accurate Positioning Algorithm/Approach

An accurate positioning approach is crucial for effective indoor localization. Chan *et al.*
(2009d) discuss applications of the Newton Trust-Region method and the use of the convergence of a trajectory to remove the noise from the received signal strength. Chan *et al.* (2010)
apply a Newton Trust-Region (TR) algorithm to trajectory estimation based on the traditional Location Fingerprinting/Localization approach. The Newton Trust-Region method
optimizes Location Fingerprinting iteratively because each point in a trajectory normally
falls into a region and converges in the same direction. More details will be discussed in
chapter 3 and 4.

1.2.2   A Stable WLAN Signal Transmission

A stable WLAN signal transmission is usually obtained by reducing the signal interference.
Interference is a major source of signal fluctuation so it is necessary to reduce channel interference between access points. Channel interference can occur because adjacent channels
overlap in the frequency spectrum in IEEE 802.11b/g WLAN. Recent research has focused
on either creating a new channel assignment scheme (Chiu *et al.*, 2006; Haidar *et al.*, 2008;
Subramanian *et al.*, 2008) or enhancing existing MAC protocols (Dutta *et al.*, 2007; Zeng
et al., 2007) to reduce channel interference. The objective of this recent work has been to
improve the data transmission through wireless networks. However, none of the existing
work on channel interference has addressed accurate performance and the improvement
of location estimation algorithms. In some cases it has even been wrongly suggested that
interference might increase the accuracy of positioning.

Chan *et al.* (2010) investigate the influence of channel interference in a location fingerprinting approach and describe localization experiments and simulations on the IEEE 802.11
test-bed. Chan *et al.* (2010) also investigate AP channel assignment, the distribution of
received signal strength (RSS) values, and variations in covered areas and the distances

between APs. This analysis will provide guidance as to (1) how to assign channels, (2) space APs to reduce interference, and (3) calculate how many access points are required to uniquely identify a location at a given accuracy and precision. The relevant findings should be of assistance to engineers in designing and better understanding a WLAN channel assignment specifically for positioning. More details will be discussed in Chapter 2.

1.2.3   A Structural WLAN Infrastructure

A structural WLAN infrastructure deployment aims to achieve high localization accuracy and optimal coverage. As alternatives, Chan *et al.* (2009c) propose three structured approaches (triangular, square and hexagonal AP distributions) to achieve this aim. Overall, the hexagonal approach is the most effective for localization operations. A surprising result of this work is the discovery that the center of a square distribution produces a localization accuracy that is 30% higher than either of the other two structured approaches. The worst performer is the ad-hoc distribution, which requires 50% more APs than the hexagonal distribution to achieve effective localization. More details will be also covered in Chapter 2.

1.2.4   A Graphical Fuzzy Signal Visualization Model

Fuzzy logic modeling can be applied to evaluate the behavior of the received signal strength (RSS) in Wireless Local Area Networks (WLANs), which is a central part of WLAN tracking analysis. Previous analytical models have considered in depth how the WLAN infrastructure affects the accuracy of tracking. Chan *et al.* (2009b) propose a novel fuzzy spatio-temporal topographic model implemented in a large (9.34 hectare), physical university campus where the WLAN received signal strength (RSS) was surveyed from more than 2000 access points. The Nelder-Mead (NM) method was applied to simplify this authors' previous work on fuzzy color maps into a topographic (line-based) map. The new model provides a detailed, quantitative representation of WLAN RSS. More details will be discussed in Chapter 6.

1.2.5   A Location-aware Information Retrieval System

One of the most challenging problems in retrieving location-aware information is to understand user queries in such a way that it is optimally suitable to the user's current location. The speed and accuracy of retrieval and the usefulness of the retrieved data depends on a number of factors including constant or frequent changes in its content or status, the effects of environmental factors such as the weather and traffic and the techniques that are used to categorize the relevance of the retrieved data.

Chan *et al.* (2009a) preliminarily attempt to deal with this task that makes use of artificial intelligent (AI) technologies operating in both wired and wireless networks to find and retrieve location-aware information. AI computer programs are endowed with a wide range of human-like abilities, including perception, the use of natural language, learning, and the ability to understand user queries. More specifically, this work proposes semantic TFIDF, an agent-based system for retrieving location-aware information that improves the speed of retrieval while maintaining or even improving the accuracy by making use of semantic information in the data to develop smaller training sets. The method assigns intelligent computer programs to first gather location-aware data and then, using semantic graphs from the WordNet English dictionary, the agents classify, match and organize the information to find a best match for a user query. Experiments show it to be significantly faster and more accurate. More details will be discussed in Chapters 11 and 12.

Chapter Summary

This chapter has introduced the background, open problems and five factors that affect the effectiveness of positioning systems. The next chapter focuses on how to build a cost-effective and resource-efficient Wi-Fi infrastructure for positioning.

References

Bahl, P. and Padmanabhan, V. (2000). RADAR: an in-building RF-based user location and tracking system. *The 19th IEEE International Conference on Computer Communications, INFOCOM*, pp. 775–784.

Budianu, C., Ben-David, S., and Tong, L. (2006). Estimation of the number of operating sensors in large-scale sensor networks with mobile access. *IEEE Transactions on Signal Processing*, pp. 1703–1715.

Chan, C. L., Baciu, G., and Mak, S. (2009a). Cognitive location-aware information retrieval by agent-based semantic matching. *The 8th IEEE International Conference on Cognitive Informatics, ICCI*, pp. 435–440.

Chan, C. L., Baciu, G., and Mak, S. (2009b). Fuzzy topographic modeling in WLAN tracking analysis. *International Conference on Fuzzy Computation, ICFC*, pp. 17–24.

Chan, C. L., Baciu, G., and Mak, S. (2009c). Resource-effective and accurate WLAN infrastructure design and localization using a cell-structure framework. *The 5th International Conference on Wireless Communications, Network and Mobile Computing*, 6: 9–15.

Chan, C. L., Baciu, G., and Mak, S. (2009d). Using the Newton Trust-Region method to localize in WLAN environment. *The 5th IEEE International Conference on Wireless and Mobile Computing, Networking and Communications, WiMOB*, pp. 363–369.

Chan, C. L., Baciu, G., and Mak, S. (2010). Properties of channel interference for WLAN location fingerprinting. *Journal of Communications Software and Systems*, 6(2): 56–64.

Chang, K., Hong, L., and Wan, H. (2007). Stochastic trust region gradient-free method (strong): a new response-surface-based algorithm in simulation optimization. *The 39th Conference on Winter Simulation*, pp. 346–354.

Chen, W., Hou, J., and Sha, L. (2004). Dynamic clustering for acoustic target tracking in wireless sensor networks. *IEEE Transactions on Mobile Computing*, pp. 258–271.

Chiu, H. S., Yeung, K., and Lui, K.-S. (2006). J-CAR: an efficient channel assignment and routing protocol for multi-channel multi-interface mobile ad hoc networks. *IEEE Global Telecommunications Conference*, pp. 1–5.

Danesh, A., Moshiri, B., and Fatemi, O. (2007). Improve text classification accuracy based on classifier fusion methods. *The 10th International Conference on Information Fusion*, pp. 1–6.

Dil, B., Dulman, S., and Havinga, P. (2006). Rangebased localization in mobile sensor networks. *European Conference on Wireless Sensor Networks*, pp. 164–179.

Dutta, P., Jaiswal, S., and Rastogi, R. (2007). Routing and channel allocation in rural wireless mesh networks. *The 26th IEEE International Conference, INFOCOM*, pp. 598–606.

Fang, S., Lin, T., and Lin, P. (2008). Location fingerprinting in a decorrelated space. *IEEE Transactions on Knowledge and Data Engineering*, 20(5): 685–691.

Haidar, M., Ghimire, R., Al-Rizzo, H., Akl, R., and Chan, Y. (2008). Channel assignment in an IEEE 802.11 WLAN based on signal-to-interference ratio. *Electrical and Computer Engineering Canadian Conference, CCECE*, pp. 001169–001174.

Hu, L. and Evans, D. (2004). Localization for mobile sensor networks. *The 10th International Conference on Mobile Computing and Networking*, pp. 45–57.

Jan, R. and Lee, Y. (2003). An indoor geolocation system for wireless LANs. *International Conference on Parallel Processing Workshops*, pp. 29–34.

Kaemarungsi, K. and Krishnamurthy, P. (2004). Modeling of indoor positioning systems based on location fingerprinting. *The 23th IEEE International Conference on Computer Communications, INFOCOM*, 2, pp. 1012–1022.

Kjaergaard, M. B. and Munk, C. V. (2008). Hyperbolic location fingerprinting: a calibration-free solution for handling differences in signal strength. *The 6th IEEE International Conference on Pervasive Computing and Communications*, pp. 110–116.

Kwon, J., Dundar, B., and Varaiya, P. (2004). Hybrid algorithm for indoor positioning using wireless LAN. *The IEEE 60th Vehicular Technology Conference, VTC*, 7, pp. 4625–4629.

Li, B., Wang, Y., Lee, H., Dempster, A., and Rizos, C. (2005). Method for yielding a database of location fingerprints in WLAN. *IEEE Proceedings on Communications*, 152(5): 580–586.

Liu, T. and Chen, H. (2004). Real-time tracking using Trust-Region methods. *IEEE Transactions on Pattern Analysis on Machine Intelligence*, pp. 397–402.

Prasithsangaree, P., Krishnamurthy, P., and Chrysanthis, P. (2002). On indoor position location with wireless LANs. *The 13th IEEE International Symposium, Personal, Indoor and Mobile Radio Communications*, 2, pp. 720–724.

Stoleru, R., He, T., Stankovic, J. A., and Luebke, D. (2005). A high-accuracy, low-cost localization system for wireless sensor networks. *The 3rd International Conference on Embedded Networked Sensor Systems*, pp. 13–26.

Stoleru, R., Vicaire, P., He, T., and Stankovic, J. (2006). StarDust: a flexible architecture for passive localization in wireless sensor networks. *The 4th International Conference on Embedded Networked Sensor Systems*, pp. 57–70.

Subramanian, A. P., Gupta, H., Das, S. R., and Cao, J. (2008). Minimum interference channel assignment in multiradio wireless mesh networks. *IEEE Transactions on Mobile Computing*, 2: 1459–1473.

Swangmuang, N. and Krishnamurthy, P. (2008). Location fingerprint analyses toward efficient indoor positioning. *The 6th IEEE International Conference on Pervasive Computing and Communications*, pp. 101–109.

Taheri, A., Singh, A., and Emmanuel, A. (2004). Location fingerprinting on infrastructure 802.11 wireless local area networks (WLANs) using Locus. *The 29th IEEE International Conference on Local Computer Networks*, pp. 676–683.

Tiemann, C., Martin, S., Joseph, J., and Mobley, R. (2000). Aerial and acoustic marine mammal detection and localization on navy ranges. *U.S. Department of Commerce and Department of the Navy, Joint Interim Report: Bahamas Marine Mammal Stranding Event*.

Wang, J., Zha, H., and Cipolla, R. (2006). Coarse-to-fine vision-based localization by indexing scale-invariant features. *IEEE Transactions on Systems, Man, and Cybernetics – Part B: Cybernetics*, 36(2): 413.

Weiss, S., Kasif, S., and Brill, E. (1996). Text Classification in USENET newsgroup: A progress report. *AAAI Spring Symposium on Machine Learning in Information Access Technical Papers, Palo Alto*, pp. 125–127.

Zeng, G., Wang, B., Ding, Y., Xiao, L., and Mutka, M. (2007). Multicast algorithms for multi-channel wireless mesh networks. *IEEE International Conference on Network Protocols, ICNP*, pp. 1–10.

Part I

Wi-Fi Positioning Systems

Chapter 2

Installation of Wi-Fi Infrastructure

Creativity is just connecting things. When you ask creative people how they did something, they feel a little guilty because they didn't really do it, they just saw something. It seemed obvious to them after a while.

Steve Jobs
1955–2011

Chapter Contents

► Choosing suitable Wi-Fi access points/routers to build the Wi-Fi positioning infrastructure

► Learning the properties of Wi-Fi signal and the channel interference

► Determining the number of access points to be installed

► Configuring the access point optimally to reduce signal interferences

► Placing the access point structurally to achieve the best positioning result

Now a days, many large shopping mall and office buildings may have many thousands of Wi-Fi access points (routers) installed to provide a web surfing service. The installation of access points is usually in an empirical and ad-hoc manner. For example, you can easily buy a wireless router and place it in your office desk to have a Wi-Fi access. Your colleagues may find that they still cannot receive a good Wi-Fi signal. They may similarly buy another wireless routers and install them. This kind of 'ad-hoc' installation cannot promise to be the magical silver bullet that can ensure no further blind spots. And this kind of 'ad-hoc' installation is very often happened in home and office environments which may lead to signal overlaps or interference because there may be too many access points packed closely together.

Introduction to Wireless Localization: With iPhone SDK Examples, First Edition. Eddie C.L. Chan and George Baciu.
© 2012 John Wiley & Sons Singapore Pte. Ltd. Published 2012 by John Wiley & Sons Singapore Pte. Ltd.

If we want to use the Wi-Fi signal strength to locate a person or a device, first we need to think about configuration attributes, such as signaling standard, transmit power, range and antenna type. These configurations affect your positioning system. Therefore, use them to advantage when tweaking the Wi-Fi positioning infrastructure.

In this chapter we'll step through the process of designing and installing the Wi-Fi positioning infrastructure. We try to answer several questions in this chapter: Which type of access point should we buy? Is there any special configuration of the Wi-Fi access point? How many access points should we install? Where should we place them?

2.1   What is the IEEE 802.11 Family?

The IEEE 802.11 family is created and maintained by the Institute of Electrical and Electronics Engineers (IEEE). The 802.11 family consists of a series of over-the-air modulation techniques that use the same basic protocol.

There are many choices of wireless local area network (WLAN) access points in the market for home and business users. Many products conform to the IEEE 802.11a, 802.11b, 802.11g and 802.11n wireless standards collectively known as Wi-Fi technologies. The most popular ones are those defined by the 802.11b and 802.11g protocols, which are amendments to the original standard. Let's briefly compare and contrast each of the wireless standards.

802.11a

802.11a supports bandwidth up to 54 Mbps and signals in a regulated frequency spectrum around 5 GHz. It has fast maximum speed with regulated frequencies preventing signal interference from other devices. This higher frequency compared to 802.11b shortens the range of 802.11a networks. The higher frequency also means 802.11a signals have more difficulty penetrating walls and other obstructions.

802.11b

802.11b access points generally offer greater range than 802.11a, mainly because 802.11b operates using lower frequencies (2.4 GHz instead of 5 GHz band). It has slower maximum speed. The production cost of 802.11b access points is lower than 802.11a. 802.11b access points may incur interference from microwave ovens, cordless phones, and other appliances using the unregulated frequency band.

802.11g

802.11g emerged on the market in 2002 and 2003. 802.11g attempts to combine the best of both 802.11a and 802.11b. 802.11g supports bandwidth up to 54 Mbps, and it uses the 2.4 Ghz frequency for greater range. 802.11g is backwards compatible with 802.11b, meaning that 802.11g access points will work with 802.11b wireless network adapters and vice versa.

802.11n

802.11n is an amendment which improves upon the previous 802.11 standards by adding multiple-input multiple-output antennas (MIMO). 802.11n operates on both the 2.4 GHz and the lesser used 5 GHz bands. The IEEE has approved the amendment and it was published in October 2009. It has the fastest maximum speed and best signal range and it is more resistant to signal interference from outside sources. 802.11n connections should support data rates of over 100 Mbps. 802.11n also offers somewhat better range over earlier Wi-Fi standards due to its increased signal intensity. 802.11n equipment will be backward compatible with 802.11g gear.

We have covered the most commonly used IEEE 802.11 networks. As the basis for making decisions about which type of access point should be purchased, take into account the coverage ranges. Generally, 802.11b and 802.11g routers/access points support a range of up to 125 feet (38 m) indoors and 460 feet (140m) outdoors. 802.11n has the largest range up to 820 feet (250 m). The effective range of 802.11a in indoor is only 115 feet (35 m) which is approximately one-third that of 802.11b/g. Table 2.1 shows the summary of 802.11 protocol.

The smaller range can have a precise positioning result, but it requires more access points to be installed to have enough coverage. The larger range means we can buy fewer access points to cover an area but we will have a relatively rough positioning. The best solution will be the mixture of these two types of access point. We have a rough positioning using the large range access points and in some places where the precise positioning is necessary, it is recommended to install small range access points and overlap the coverage.

2.2 Properties of Wi-Fi Signal Strength

Understanding the properties of Wi-Fi signal strength is important for the design of positioning systems. The analysis of Wi-Fi signal strength provides insights into how to place APs so as to reduce interference, and how many access points are required to uniquely identify a location at a given accuracy and precision. The analysis of Wi-Fi signal strength would be of interest and assistance to engineers designing Wi-Fi infrastructure specifically for positioning.

In this section, we will look at the distribution of Wi-Fi received signal strength (RSS) values, their standard deviations, their temporal variations, and the (in)dependence of received signal strengths (RSSs) from multiple access points (APs).

2.2.1 Distribution of Wi-Fi Signal Strength

The distribution of Wi-Fi signal strength is highly dependent on the signal propagation effects such as reflection, diffraction, and scattering. The dense multi-path environment causes the received Wi-Fi signal to fluctuate around a mean value. Wi-Fi signal propagation loss is governed by:

$$R = r - 10\alpha \log_{10}(d) - wallLoss \tag{2.1}$$

where r is initial received signal strength, d is a distance from access points (APs) to a location, α is the path loss exponent (clutter density factor) and $wallLoss$ is the sum of the losses introduced by each wall on the line segment drawn at Euclidean distance d.

Initially, r is the initial RSS at the reference distance of d_0 is 1 meter (this is 41.5 dBm for LOS propagation and for 37.3 dBm non-line-of-sight (NLOS) propagation for some report

Table 2.1 Summary of 802.11 Protocol.

802.11 Protocol	Release Date	Frequency (GHz)	Bandwidth (MHz)	Data rate (Mbit/s)	Modulation	Approximate Indoor Range		Approximate Outdoor Range	
						(m)	(ft)	(m)	(ft)
802.11 a	Sep 1999	5	20	6, 9, 12, 18, 24, 36, 48, 54	OFDM	35	115	120	390
802.11 b	Sep 1999	2.4	20	5.5, 11	DSSS	38	125	140	460
802.11 g	Jun 2003	2.4	20	6, 9, 12, 18, 24, 36, 48, 54	OFDM, DSSS	38	125	140	460
802.11 n	Oct 2009	2.4/5	20	7.2, 14.4, 21.7, 28.9, 43.3, 57.8, 65, 72.2	OFDM	70	230	250	820
			40	15, 30, 45, 60, 90, 120, 135, 150					

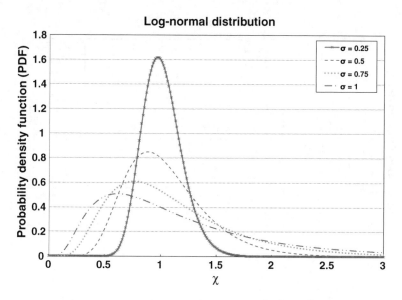

Figure 2.1   Log-normal distribution (large value of path loss).

measurement). The path loss exponent α at a carrier frequency of 2.4 GHz is reported to be 2 for line-of-sight (LOS) propagation and 3.3 for NLOS propagation. Under other circumstances, α can be between 1 and 6. It is not easy to have an accurate positioning if the environment has moving objects and numerous reflecting surfaces. This leads to the severe multi-path problem which implies low probability for availability of line-of-sight path.

2.2.2   Large Value of Path Loss

The large value of path loss exponent α describes the signal attenuation as the signal travels over a distance and is absorbed by materials such as walls and floors along the way to the receiver. This component predicts the mean of the RSS and usually has a log-normal distribution (Kaemarungsi and Krishnamurthy, 2004) :

$$f(x\,|\mu,\sigma) = \frac{1}{x\sqrt{2\pi\sigma^2}}e^{\frac{-(\ln x-\mu)^2}{2\sigma^2}} \tag{2.2}$$

where μ and σ are the mean and standard deviation. Figure 2.1 shows the log-normal distribution for the large value of path loss.

2.2.3   Small Value of Path Loss

The small value of path loss exponent α explains the dramatic fluctuation of the signal due to multi-path fading. If there is no line-of-sight (NLOS) component, the small-scale fading is often modeled with a Rayleigh distribution (Kaemarungsi and Krishnamurthy, 2004):

$$f(x;\sigma) = \frac{x}{\sigma^2}e^{-x^2/2\sigma^2} \tag{2.3}$$

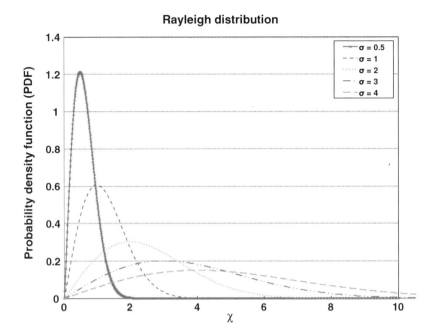

Figure 2.2   Rayleigh distribution (small value of path loss).

If there is a line-of-sight (LOS) component, the small-scale fading is modeled with a Rice distribution.

$$f(x \mid v, \sigma) = \frac{x}{\sigma^2} e^{-(x^2 + v^2)/2\sigma^2} I_0 \left(\frac{xv}{\sigma^2}\right) \qquad (2.4)$$

where $I_0(z)$ is the modified Bessel function of the first kind with order zero. For simplicity, we will take $v = 0$, the distribution reduces to a Rayleigh distribution. Figure 2.2 shows the Rayleigh distribution for the small value of path loss.

2.2.4   Behavior Study on the Human's Presence

The human body consists of 70% water. Water absorbs the signal strength more than the air. The effect of the user's presence can affect the mean of the received signal strength (RSS) value, and then it could affect the accuracy of location estimation in the system.

To measure the effect of the human body, a simple experiment was performed to measure the received signal strength (RSS) from an access point (AP) using a receiver. The distance between the AP and the receiver is approximately 5 meters and one concrete wall partition was placed between them. The sampling schedule is to collect the RSS data every 5 seconds within 2 hours. The results were analyzed by two plotting histograms of the RSS. Figure 2.3 shows the histogram of RSS with the presence of a human. Figure 2.4 shows the histogram of RSS with the absence of human.

Clearly, the human body influences the RSS distribution by spreading the range of RSS values by a significant amount. When the positioning system is supposed to cater to real users, it is essential to have the user present while collecting the RSS values for the fingerprint

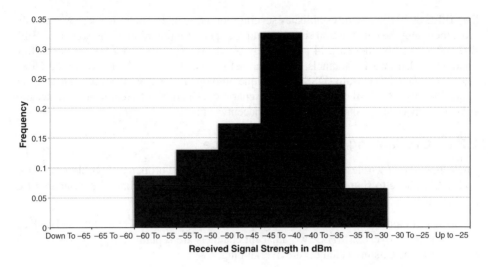

Figure 2.3   Histogram of RSS with the presence of a human.

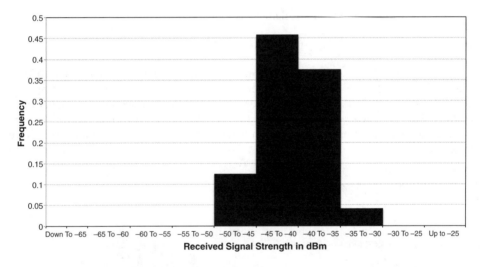

Figure 2.4   Histogram of RSS with the absence of a human.

and to take into account the effect of the human body. For applications that make use of sensors without a human presence the data should reflect that environment (Kaemarungsi and Krishnamurthy, 2004).

2.3 Optimal Channel Allocation for Wi-Fi Positioning

In this section, we talk about how to achieve the optimal channel allocation for the Wi-Fi positioning. Before this, it is essential to know how the IEEE 802.11 standard establishes different channels and to study the properties of channel interference.

The IEEE 802.11 standard establishes several requirements for radio frequency transmission, including the canalization schemes and the spectrum radiation of the signal. In IEEE 802.11 b/g WLAN, there are 14 channels. In North America, the 2.4 GHz frequency ISM band is divided into 11 channels. Each channel is spread over 22 MHz due to the Direct Sequence Spread Spectrum (DSSS) technique employed by IEEE 802.11b/g. These channels have only 5 MHz of center frequency separation. Channel interference occurs because frequency spectrum is shared with each adjacent channel.

2.3.1   Overlapping Channel Interference

The bandwidth of wireless network is limited because of the property of wireless networks and stations have to share the limited bandwidth. IEEE 802.11b/g has 14 overlapping frequency channels. Channel 1, 6 and 11 are non-overlapping channels.

As shown in Figure 2.5, IEEE 802.11 b/g spreads through 2401 MHz to 2483 MHz. Each channel spreads over 22 MHz. Two adjacent channels are separated only by 5 MHz such that most of the existing channels are overlapping.

Interference-level Function

The interference-level function γ is defined as follows:

$$\gamma(\Delta c) = \max(0, 1 - k\Delta c) \qquad (2.5)$$

where Δc is the absolute channel difference and k is the non-overlapping ratio of two channels. γ and Δc are in Db unit. When Δc increases, γ decreases. For example, if $\Delta c = 0$, then $\gamma(\Delta c) = 1$ and if $\Delta c \geq 5$, then $\gamma(\Delta c) = 0$. In other words, for channel 1 and 6, $\Delta c = 5$, $k = 0$, then $\gamma(\Delta c) = 0$, suggesting no interference. In a real case, if APs are installed and placed far enough away from others, γ should be at least equal to the above threshold.

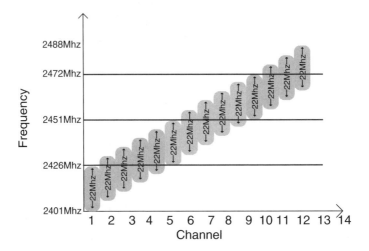

Figure 2.5   IEEE 802.11b/g frequency spectrum to channel number.

Signal-to-Interference-plus-Noise-Ratio

Signal-to-Interference-plus-Noise-Ratio (SINR) is a very common indicator to measure interference. SINR is defined as follows:

$$SINR = \frac{R_b}{\gamma(\Delta c) \sum R + n} \tag{2.6}$$

where R_b is the highest RSS after path loss calculation. R is the remaining set of RSS after path loss calculation. n is the noise signal strength. R_b, R, n are in dBm unit. Usually, n should have the value of -100 dBm. Again SINR should be at least equal to above calculated threshold which depends on the distance among APs, the transmission rate, the modulation scheme and the required bit-error rate. The higher the value of SINR is, the more accurate the positioning is.

2.3.2 Distribution of Channel Interference

The signal difference between the preset (maximum) RSS value of AP and sample RSS of an AP could be seen as interference strength value (in dB). Two APs are placed and interfere with each other. We assume that there is a zero (or very short) distance between receiver and APs, In other words, the signal should not be reduced by propagation-loss. The preset RSS value is denoted by ρ and the interference strength value by $F = \{f_1, f_2, ..., f_n\}$. The interference could then be by:

$$Y = \sqrt{\sum_{i=1}^{n} (\rho - r_i)^2} = \sqrt{\sum_{i=1}^{n} f_i^2} \tag{2.7}$$

where n represents the number of collected sample. In (2.7), the difference between a preset RSS value and measured RSS, r_i, is considered to be a signal interference strength.

We have discussed about the distribution of the RSS and it is often left-skewed. Therefore, the random variable of interference f_i has a nonzero mean and usually obeys a noncentral chi-square distribution:

$$P_X(x; n, \lambda) = \frac{1}{2\sigma^2} e^{-\frac{x+\lambda}{2\sigma^2}} \left(\frac{x}{\lambda}\right)^{\frac{n-2}{4}} I_{\frac{n-2}{2}} \left(\frac{\sqrt{\lambda x}}{\sigma^2}\right) \tag{2.8}$$

where $I_k(x)$ is the kth-order modified Bessel function of the first kind given by

$$I_k(x) = \left(\frac{x}{2}\right)^k \sum_{i=0}^{\infty} \frac{x^{2i}}{4^i i! \Gamma(k+i+1)} \tag{2.9}$$

To measure the channel difference, a simple experiment was performed. In the experiment two APs were placed within a short distance of a RSS receiver. Two APs were set to the same channel and emitted a WLAN signal at the maximum strength. The assumption is that all signal fluctuations are caused by interference between two access points (APs). A receiver recorded samples of signal strength from two APs over two hours. The sample result was used to plot Figures 2.6 and 2.7.

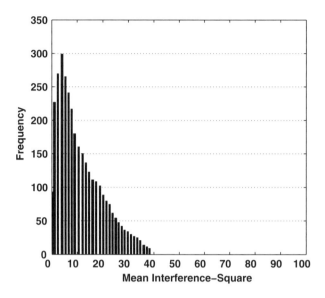

Figure 2.6   Result of frequency to mean interference-square experiment.

Figure 2.6 shows the relationship of the mean interference-square to frequency. Along the Y-axis, frequency represents the number of occurrences of a particular value of the interference-square. Figure 2.7 shows the relationship of the mean interference-square to the probability density function (pdf).

We will now discuss the impact of parameters of the interference based on the visual results (Figures 2.6 and 2.7). The visual results so far suggest that interference mostly occurs in a right-skewed distribution with a long tail to infinity. A smaller value for the

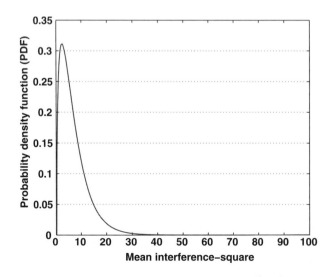

Figure 2.7   Result of PDF to mean interference-square experiment.

mean interference-square has a larger value of the pdf. It also shows that a noncentral chi-distribution could represent the distribution of interference. It would be ideal to have less interference that implies a smaller fluctuation. The above result helps to understand the features of interference and it could be used to decide a better positioning system.

2.3.3 Channel Assignment Schemes

There are typically three channel assignment schemes: ascending, orthogonal and ad-hoc channel allocations. Ascending channel allocation assigns the channel of each neighbor AP in ascending order (1, 2, 3 ...). Ad-hoc channel allocation randomly assigns the channel of each AP from channel 1 to channel 14. Orthogonal channel allocation assigns the channel of each neighbor AP not to overlay the other channels' spectrum, for example, channel 1, 6 and 11.

Channel assignment is critical issue to positioning system. Choosing orthogonal channel allocation could save 15% of APs and meanwhile achieve 10% more accurate positioning in average. This is because orthogonal channel allocation provides less interference than the other two channel allocations.

An experiment was performed to see the positioning accuracy relationship of number of access points using these three channel assignments. Figure 2.8 shows the relationship of the number of access points to accuracy using each of the three allocation schemes. The target resolution is 2 m and the experiment was conducted in a 1000 sq. meter lab area. A higher number of APs improves the precision dramatically up to the point that nine APs are used. If more than nine APs are used, the accuracy does not increase significantly due to the interference between them. The result shows that orthogonal channel allocation with only nine APs achieve 90% accuracy. Perhaps the most important point to note is that orthogonal channel allocation could require 15% fewer APs than either ascending channel allocation or ad-hoc channel allocation.

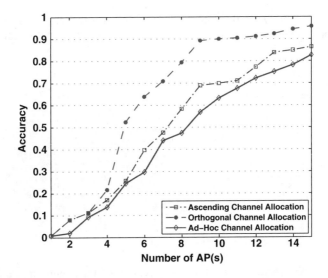

Figure 2.8   Relationship of number of access points to accuracy under ascending, orthogonal and ad hoc channel allocation.

A higher number of APs gives more accurate positioning. Moreover, as the number of APs increases, the channel interferences among APs increase and decrease the rate of increasing positioning accuracy. Reducing channel interference is essential to improve the positioning accuracy and meanwhile could save significant amount of resources of localization infrastructure.

2.4   Determining Number of APs to be Installed

In this section we figure out how many access points are needed to achieve the best positioning. First, we introduce square tessellation installation approach. Then we define what the Z factor is and the environmental factors that cause signal loss. Finally, we make use of the Z factor, the relationship between building type and positioning resolution to estimate how many access points are needed to be installed in the square tessellation.

2.4.1   Square Tessellation Installation

Square tessellation installation is a common and standard way to deploy the AP evenly in square cells. Each AP is installed at the center of the square cell. It is often used as an initial deployment plan in practice. Figure 2.9 shows the square tessellation deployment where AP is located at center and we assume that the access point emits RF signal evenly in any direction which has a circular coverage.

2.4.2   Z Factor

Z factor is the length of a square that corresponds to the coverage area of the access point. Figure 2.10 illustrates the idea of radius and z factor. Table 2.2 shows the calculated result of coverage area, radius and Z factor.

2.4.3   Environmental Factors

Environmental operability should be considered as well. These factors include temperature, moisture, human movements, placement of furniture and thickness of the wall. For instance,

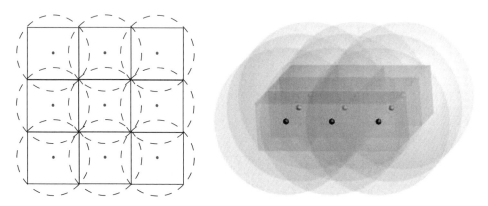

Figure 2.9   Square tessellation. (A full color version of this figure appears in the color plate section.)

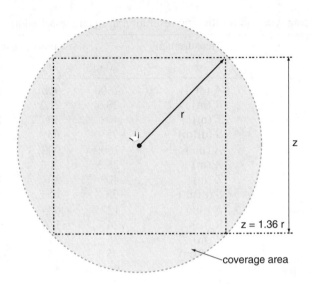

Figure 2.10   Coverage radius r and z factor.

if the positioning system is operated in a busy environment with lots of furniture, it usually requires more APs to locate the device precisely.

The relationship between the building type and positioning resolution are summarized in Table 2.3. N meters resolution means we can locate a device correctly in a precision of N meters. Office space with brick walls is usually required to install many APs and pack them closely. If the positioning resolution is target for 3 m, each AP should be installed 5m apart with a strong signal strength, -30 dBm.

2.4.4   Number of Access Points Needed

We first need to get the floor plan layout and then divide it into rectangle partitions. With Table 2.3 and given positioning resolution, we can estimate how many APs are required to be installed. The example shown in Figure 2.11 is for a typical office building with a desired position resolution of 5 meters.

Table 2.2   Radius and Z factor table in square tessellation.

Area (square m)	r (m)	z (m)
1600	28.3	40
1400	26.5	37.4
1200	24.5	34.6
1000	22.4	31.6
800	20	28.3
600	17.3	24.5
400	14.1	20
200	10	14.1

Table 2.3   Building type and positioning resolution in square tessellation.

Building type	Measurement	Positioning resolution			
		20 m	10 m	5 m	3 m
Home	A (m^2)	1600	900	225	100
	R (m)	28.3	21.2	10.6	7.1
	Z (m)	40	30	15	10
	S (dBm)	−75	−65	−40	−30
Typical office	A (m^2)	1600	900	225	100
	R (m)	28.3	21.2	10.6	7.1
	Z (m)	40	30	15	10
	S (dBm)	−75	−65	−40	−30
Drywall office space	A (m^2)	1225	625	144	49
	R (m)	24.7	17.7	8.5	4.9
	Z (m)	35	25	12	7
	S (dBm)	−75	−65	−40	−30
Brick wall office space	A (m^2)	900	400	100	25
	R (m)	21.2	14.1	7.1	3.5
	Z (m)	30	20	10	5
	S (dBm)	−75	−65	−40	−30
Warehouse/manufacturing	A (m^2)	1600	900	225	100
	R (m)	28.3	21.2	10.6	7.1
	Z (m)	40	30	15	10
	S (dBm)	−75	−65	−40	−30

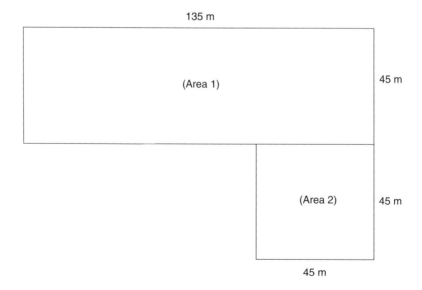

Figure 2.11   L-shape floor plan.

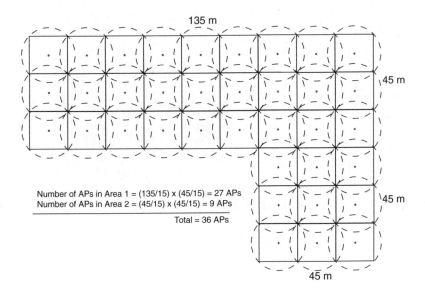

Figure 2.12 Access points location.

In Table 2.3, we can find that Z factor of a typical office environment is 15 m. The total office area is 8100 m^2. Around 36 APs are estimated to be installed. The calculation is also shown in Figure 2.12. After the calculation, we use the radius or Z factor found in Table 2.3, distribute the circles or squares evenly across the rectangles found. Notice that the squares are directly adjacent to each other, as shown in Figure 2.12.

Square tessellation is usually adopted in practice but it is not the most effective one. In the next section, we introduce two other tessellations, triangular and hexagonal.

2.5 Other Tessellation Installations

In this section, we will introduce two tessellations, triangular and hexagonal. Similarly, access points (APs) are installed and tessellated in triangular and hexagonal distributions. Each AP is installed at the center of the cell. Figures 2.13 and 2.14 show the triangular and hexagonal tessellation deployment where AP is located at the center.

We now define what X and Y factors are and make use of X, Y factors and the relationship between building type and positioning resolution to estimate how many access points are needed to be installed in the square tessellation.

2.5.1 X and Y Factors

X factor is the horizontal distance to the next access point. Y factor is the vertical distance of a cell that corresponds to the coverage area of the access point. Figures 2.15 and 2.16 illustrate the idea of radius, X and Y factors. The resulting table of radius, X and Y factor is shown next to each of the two figures.

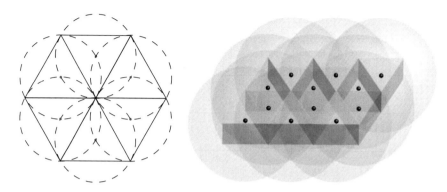

Figure 2.13   Triangular tessellation. (A full color version of this figure appears in the color plate section.)

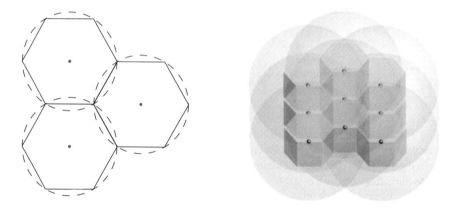

Figure 2.14   Hexagonal tessellation. (A full color version of this figure appears in the color plate section.)

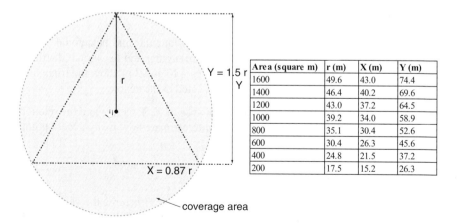

Area (square m)	r (m)	X (m)	Y (m)
1600	49.6	43.0	74.4
1400	46.4	40.2	69.6
1200	43.0	37.2	64.5
1000	39.2	34.0	58.9
800	35.1	30.4	52.6
600	30.4	26.3	45.6
400	24.8	21.5	37.2
200	17.5	15.2	26.3

Figure 2.15   Coverage radius r, X and Y factors in the triangular tessellation.

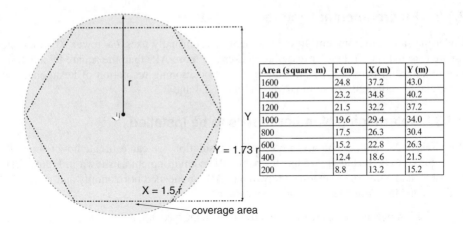

Area (square m)	r (m)	X (m)	Y (m)
1600	24.8	37.2	43.0
1400	23.2	34.8	40.2
1200	21.5	32.2	37.2
1000	19.6	29.4	34.0
800	17.5	26.3	30.4
600	15.2	22.8	26.3
400	12.4	18.6	21.5
200	8.8	13.2	15.2

Figure 2.16   Coverage radius r, X and Y factors in the hexagonal tessellation.

Table 2.4   Building type and positioning resolution in the hexagonal tessellation.

Building type	Measurement	Positioning resolution			
		20 m	10 m	5 m	3 m
Home	A (m^2)	2081	1168	292	131
	R (m)	28.3	21.2	10.6	7.1
	X (m)	42.5	31.8	15.9	10.7
	Y (m)	49.0	36.7	18.4	12.3
	S (dBm)	−75	−65	−40	−30
Typical office	A (m^2)	2081	1168	292	131
	R (m)	28.3	21.2	10.6	7.1
	X (m)	42.5	31.8	15.9	10.7
	Y (m)	49.0	36.7	18.4	12.3
	S (dBm)	−75	−65	−40	−30
Drywall office	A (m^2)	1585	814	188	62
	R (m)	24.7	17.7	8.5	4.9
	X (m)	37.1	26.6	12.8	7.4
	Y (m)	42.8	30.7	14.7	8.5
	S (dBm)	−75	−65	−40	−30
Brick wall office	A (m^2)	1168	517	131	32
	R (m)	21.2	14.1	7.1	3.5
	X (m)	31.8	21.2	10.7	5.3
	Y (m)	36.7	24.4	12.3	6.1
	S (dBm)	−75	−65	−40	−30
Warehouse/manufacturing	A (m^2)	2081	1168	292	131
	R (m)	28.3	21.2	10.6	7.1
	X (m)	42.5	31.8	15.9	10.7
	Y (m)	49.0	36.7	18.4	12.3
	S (dBm)	−75	−65	−40	−30

2.5.2   Environmental Factors

Under the same environment, hexagonal tessellation usually uses 7% fewer APs than the square tessellation. Triangular tessellation uses 5% more APs than the square tessellation. The relationship between the building type and positioning resolution of hexagonal and triangular tessellations are summarized in Tables 2.4 and 2.5.

2.5.3   Determining Number of APs to be Installed

With Tables 2.4 and 2.5 and given positioning resolution, we can estimate how many APs are required to be installed. We use the same office environment as shown in Figure 2.11. We use X and Y factors to estimate how many APs are needed horizontally and vertically. Let's use the hexagonal tessellation as an example.

Figure 2.17 shows how APs are distributed evenly in the hexagonal tessellation. The number of APs is also calculated in Figure 2.17. The numbers of APs in vertical and horizontal dimension are rounded to the nearest integer value. Around 22 APs are estimated to be installed in a strict implementation.

Table 2.5   Building type and positioning resolution in the triangular tessellation.

Building type	Measurement	Positioning resolution			
		20 m	10 m	5 m	3 m
Home	A (m^2)	1040	584	146	65
	R (m)	28.3	21.2	10.6	7.1
	X (m)	24.5	18.4	9.2	6.1
	Y (m)	42.5	31.8	15.9	10.7
	S (dBm)	−75	−65	−40	−30
Typical office	A (m^2)	1040	584	146	65
	R (m)	28.3	21.2	10.6	7.1
	X (m)	24.5	18.4	9.2	6.1
	Y (m)	42.5	31.8	15.9	10.7
	S (dBm)	−75	−65	−40	−30
Drywall office	A (m^2)	793	407	94	31
	R (m)	24.7	17.7	8.5	4.9
	X (m)	21.4	15.3	7.4	4.2
	Y (m)	37.1	26.6	12.8	7.4
	S (dBm)	−75	−65	−40	−30
Brick wall office	A (m^2)	584	258	65	16
	R (m)	21.2	14.1	7.1	3.5
	X (m)	18.4	12.2	6.1	3.0
	Y (m)	31.8	21.2	10.7	5.3
	S (dBm)	−75	−65	−40	−30
Warehouse/manufacturing	A (m^2)	1040	584	146	65
	R (m)	28.3	21.2	10.6	7.1
	X (m)	24.5	18.4	9.2	6.1
	Y (m)	42.5	31.8	15.9	10.7
	S (dBm)	−75	−65	−40	−30

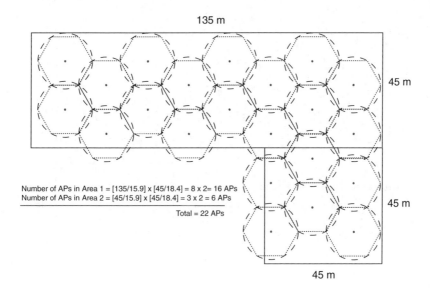

Figure 2.17   Access points location in the hexagonal tessellation.

Strict implementation of the hexagonal approach would be too ideal because architecture of a building may not allow such AP deployment. For example, the depth of walls and floors and their composition (e.g. concrete) would affect issues of AP placements.

In real world scenarios, engineers need to go through every floor of every building in a construction site to detect whether a region is sufficient enough to be covered by wireless APs. After they collect wireless signal sample, they will decide where APs should be added to increase signal coverage or be taken away to reduce signal interference. Very often, Wi-Fi signal strengths are not stable at the corner area and extra APs should be placed.

The hexagonal approach could act as blueprint of deployment to save massive labor works for signal surveying. Engineers can base their work on this blueprint and make reasonable installation adjustments. Figure 2.18 shows the final plan of the APs' location. Approximately around 33 APs should be installed to achieve 5 m resolution in the entire office space.

2.5.4   Summary of AP Deployment Approach

Overall, the hexagonal approach is by far the most cost-effective approach and engineers may wish to take account of this in their infrastructure designs and deployment. It is of interest to note, and may be useful to know in some circumstances, that where few APs are to be deployed, the triangular approach may outperform the square and hexagonal approaches.

In particular in some real world scenarios, we have the issue of limited resources to achieve the higher localization accuracy. Usually, in the case of office and exhibition hall environments, the center of localization accuracy is much more important than the side. If we just consider the center of area, the center part of localization accuracy of square distribution

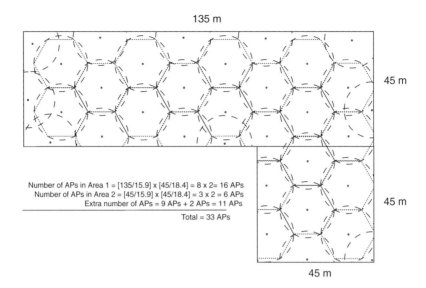

Figure 2.18   Access points location in the hexagonal tessellation (after adjustment).

is much more effective than hexagonal, which makes center of localization 10% more accurate. There is a trade-off between the localization accuracy of the centre and side part. Scarifying the side part of localization accuracy may be a possible solution to solve this issue.

Chapter Summary

This chapter has covered how to choose appropriate Wi-Fi access points and configure how to place access points optimally. The properties of Wi-Fi signal and the channel interference were studied and it was shown that when the positioning system is supposed to cater to real users, it is essential to have the user present while collecting the RSS values for the fingerprint and to take into account the effect of the human body. Finally, readers can use the content of this chapter to determine how many APs are required to achieve a certain resolution. Having established an optimized Wi-Fi infrastructure in this chapter, we will next learn the positioning algorithms.

Reference

Kaemarungsi, K. and Krishnamurthy, P. (2004). Properties of indoor received signal strength for WLAN location fingerprinting. *The 1st International Conference on Mobile and Ubiquitous Systems: Networking and Services, MOBIQUITOUS*, pp. 14–23.

Chapter 3

Algorithms Used in Wi-Fi Positioning Systems

When you first start off trying to solve a problem, the first solutions you come up with are very complex, and most people stop there. But if you keep going, and live with the problem and peel more layers of the onion off, you can often times arrive at some very elegant and simple solutions.

Steve Jobs
2006

Chapter Contents

- ► Understanding basic types of wireless localization
- ► Learning propagation-based positioning techniques
- ► Learning location-fingerprinting-based positioning techniques
- ► Introducing the fundamental algorithms for indoor positioning techniques
- ► Evaluating the performance of indoor positioning techniques
- ► Comparing different indoor positioning techniques

Indoor localization have become very popular in recent years. Wi-Fi is a wireless technology brand owned by the Wi-Fi Alliance intended to improve the interoperability of wireless local area network products based on the IEEE 802.11 standards. Wi-Fi city concepts have been adopted in many cities all over the world, such as Hong Kong, Taipei, Singapore, Tokyo, New York and many others. Users not only freely explore the Internet but also make use of existing Wi-Fi infrastructure for location-sensing in both indoor

Introduction to Wireless Localization: With iPhone SDK Examples, First Edition. Eddie C.L. Chan and George Baciu.
© 2012 John Wiley & Sons Singapore Pte. Ltd. Published 2012 by John Wiley & Sons Singapore Pte. Ltd.

and outdoor environments. Wi-Fi positioning may be a useful supplementary solution in localization.

An astonishing growth of indoor positioning systems has been made during the last decade. Especially in the area of the Wi-Fi indoor positioning system, both academia and industry are still exploring the possibility of high accuracy and precision, robust and scalable solutions. Many research and commercial products in this area are new and some open problems in research are still not solved.

3.1 Taxonomy of Indoor Positioning Techniques

Indoor positioning techniques can be categorized into three general types: scene analysis (Hightower and Borriello, 2001), triangulation (van de Goor, 2009), and proximity (Torre and Rallet, 2005). Figure 3.1 provides taxonomy of the three different types of indoor positioning techniques.

Scene Analysis

Scene analysis positioning techniques collects and extracts the features from observed scene. The observed features are usually specific and unique. The scene could be radio frequency waves, acoustic sound, visual images or any other measurable physical phenomena which usually exist near to the object. The typical technique of RF-based scene analysis is location fingerprinting (LF). The observed features (these are known as 'fingerprints' but should not be confused with the 'fingerprints' of location fingerprinting) are usually specific and unique and are used to estimate the location of the observer or of observed objects in the scene. We can estimate the distance by matching the similarity of features.

Static scene analysis searches for observed features in a predefined dataset that maps them to object locations. Differential scene analysis estimates location by tracking the difference

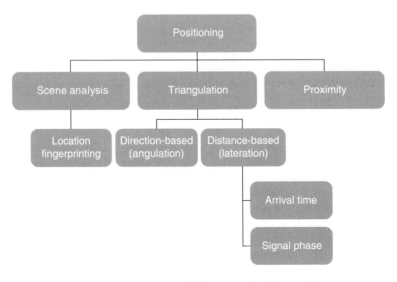

Figure 3.1   Three types of indoor positioning techniques.

between successive scenes. Differences in the scenes correspond to movements of the observer and if the features in the scene are known to be at specific positions, the observer can compute their positions (Metsala, 2004).

Triangulation

Triangulation uses the geometric properties of triangles to estimate the target location. It can be divided into the subcategories of lateration, using distance measurements, and angulations, using primarily angle or bearing measurements. Lateration estimates the position of an object by measuring its distances from multiple reference points. So, it is also called range measurement techniques. Instead of measuring the distance directly using received signal strengths (RSS), time of arrival (TOA) or time difference of arrival (TDOA) is usually measured, and the distance is derived by computing the attenuation of the emitted signal strength or by multiplying the radio signal velocity and the travel time. Roundtrip time of flight (RTOF) or received signal phase method is also used for range estimation in some systems. Angulation locates an object by computing angles relative to multiple reference points.

Proximity

The proximity technique usually provides symbolic relative location information. This technique identifies an object with a tag and a dense grid of antennas in a known location detects the tag. If the tag is detected by a single antenna, we can say that the object is 'near' to that antenna. When more than one antenna detect the mobile target, it is considered to be collocated with the one that receives the strongest signal. This method is relatively simple to implement. It can be implemented over different types of physical media. The presence of the object is sensed using a mechanism with a limited range, for example, infrared radiation (IR), radio frequency identification (RFID), pressure and touch sensors, and capacitive field detectors. Monitoring is possible when a mobile device or tag is in the range of one or more antennas.

Another example is the cell identification (Cell-ID) or cell of origin (COO) technique (Liu *et al.*, 2007). This method relies on the fact that mobile cellular networks can identify the approximate position of a mobile handset by knowing which cell site the device is using at a given time. The main benefit of Cell-ID is that it is already in use today and can be supported by all mobile handsets.

After we understand the basic categories of indoor positioning techniques, we further look at the most typical algorithms, propagation based and location fingerprinting (LF) based.

3.2 Propagation-based Algorithms

Propagation based algorithms (Prasithsangaree *et al.*, 2002; Jan and Lee, 2003) estimates the position by measuring the received signal strength with path loss. Signal propagation loss algorithm (Li *et al.*, 2005) calculates the received signal strength (RSS) with path loss as follows:

$$R = r - 10\alpha \log_{10}(d) - wallLoss \tag{3.1}$$

where r is initial RSS, d is a distance from access points (APs) to a location, α is the path loss exponent (clutter density factor) and $wallLoss$ is the sum of the losses introduced by each wall on the line segment drawn at Euclidean distance d.

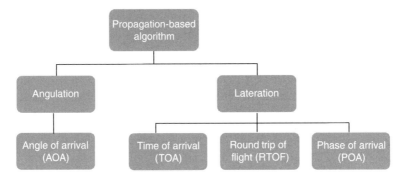

Figure 3.2   Taxonomy of propagation-based algorithms.

It is not easy to model the radio propagation in the indoor environment because of severe multi-path, low probability for availability of LOS path, and specific site parameters such as floor layout, moving objects and numerous reflecting surfaces. There is no good model for indoor radio multi-path characteristic so far (Liu *et al.*, 2007). Propagation-based techniques usually apply mathematical models to a set of triangulation algorithms to determine the position of the device. The triangulation approach uses the geometric properties of triangles to estimate the target location. It has two derivations: angulation and lateration. Angulation estimates the position of an object by computing angles relative to multiple reference points. Lateration estimates the position of an object by measuring its distances from multiple reference points. So, it is also called range measurement techniques.

For the completeness of the content, we briefly introduce some propagation-based techniques, such as the triangulation approach using angle of arrival (AOA) , phase of arrival (POA), time of arrival (TOA) and roundtrip time of flight (RTOF) techniques to determine the location of the device. Figure 3.2 shows the taxonomy of propagation-based algorithms.

3.2.1   Angle of Arrival (AOA)

Angle of arrival (AOA) method makes use of incident angles of WLAN signal to a WLAN-enabled receiver (e.g. mobile phone or notebook which can detect Wi-Fi) and the intersection of angle direction lines to estimate the position. AOA assumes the mobile device can detect the incident angle of WLAN signal or know the direction of access point that broadcast the WLAN signal. The direction (angle) of an incident WLAN signal is known as orientation. Orientation is in degrees in a clockwise direction from the North. When pointing to the North, the orientation is 0°. A single AOA measurement restricts the estimation along a line to the source. AOA requires at least two access points from two different locations; the position of the user can be estimated at the intersection of the lines of bearing from the two access points.

Figure 3.3 shows the AOA positioning method by measuring angles from two access points. As early WLAN-enabled receivers fail to detect the incident angle of a WLAN signal, instead of using incident angles ϕ_1 and ϕ_2, we can use θ_1, θ_2 because alternate angles are equal in two parallel lines. θ_1, θ_2 can be measured if we know the positions of access points. That's why AOA also is called direction of arrival (DOA). AOA methods require at least two

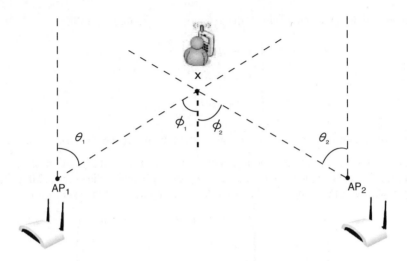

Figure 3.3   Positioning by measuring angles from two access points.

known access points AP_1, AP_2 and two measured angles θ_1, θ_2 to derive the 2D position of a user, X.

In most of the cases, triangulation approach with AOA method will be used together to estimate the position to increase the positioning accuracy. Figure 3.4 shows 2D positioning case using three access points (APs). The position of these three APs' location are denoted as (x_1, y_1), (x_2, y_2), (x_3, y_3) and the user position is (x, y).

Using the propagation loss algorithm (3.1), assume the distances between the access points and user location to be d_1, d_2, and d_3, where d_0 is the initial received signal strength at the

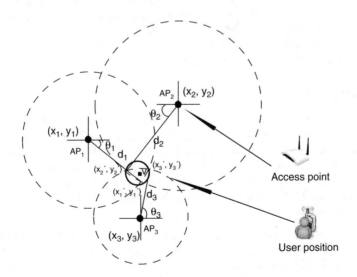

Figure 3.4   Determining 2D position of a user using AOA method.

reference distance. We can calculate the distance d_1, d_2, and d_3 as follows:

$$\begin{cases} d_1 = d_0 10^{\frac{r_0 - r_1 - wallLoss}{10 \cdot \alpha}} \\ d_2 = d_0 10^{\frac{r_0 - r_2 - wallLoss}{10 \cdot \alpha}} \\ d_3 = d_0 10^{\frac{r_0 - r_3 - wallLoss}{10 \cdot \alpha}} \end{cases} \tag{3.2}$$

Initially, r_0 is the initial received signal strength at the reference distance of d_0.

We can use the triangulation approach with AOA to estimate the position of user. After calculating the distance, we find the angle θ_1, θ_2 and θ_3 between the user position and APs, and then we are able to calculate the possible position matrix of the user as follows:

$$\begin{pmatrix} x'_1 & y'_1 \\ x'_2 & y'_2 \\ x'_3 & y'_3 \end{pmatrix} = \begin{pmatrix} x_1 + d_1 \cos \theta_1 & y_1 + d_1 \sin \theta_1 \\ x_2 + d_2 \cos \theta_2 & y_2 + d_2 \sin \theta_2 \\ x_3 + d_3 \cos \theta_3 & y_3 + d_3 \sin \theta_3 \end{pmatrix} \tag{3.3}$$

Averaging the value of position matrix, we can estimate the user position as follows:

$$x = \frac{x'_1 + x'_2 + x'_3}{3} \tag{3.4}$$

$$y = \frac{y'_1 + y'_2 + y'_3}{3} \tag{3.5}$$

There are several drawbacks to the AOA method. First, it suffers the location estimate degradation because of multi-path reflections, shadowing and direction of signal reception. Second, it assumes that there is no time delay and interference when an incident WLAN signal travels a long path from the access point (transmitter) to the mobile device (receiver). Third, it requires us to know about the configuration of WLAN infrastructure, for example, the position of APs, walls and furniture and it is so difficult to have such a blueprint of WLAN infrastructure in an ad hoc environment.

3.2.2 Time of Arrival (TOA)

Time of arrival (TOA) assumes the distance between an access point and a WLAN-enabled device which is directly proportional to signal propagation. TOA makes use of this proportional properties and triangulation approach to locate the WLAN-enabled device where the radius of three access points' signal intersects. TOA requires us to put time-stamps in the WLAN signal which helps to label when the signal travels. In order to perform an accurate estimation, TOA requires us to precisely synchronize all the access points and the WLAN-enabled device.

Range measurement is to calculate the distance between an access point and the WLAN-enabled device. The WLAN signal travels with the speed of light in air. The range measurement (ℓ) can be calculated as follows:

$$\ell_i = c(t_i - t_s) - \xi, i = 1, 2, ..., n \tag{3.6}$$

where d_i is the distance between ith access points and a WLAN-enabled device, t_i is the time when signal receives from the ith access point, t_s is the time when the WLAN signal transmits from the access points, ξ is the range error or noise, c is the speed of light and n is the total number of access points.

Let (x_i, y_i) be the coordinates of ith AP, (x, y) be the user position. The range measurement ℓ_i is:

$$\ell_i = \sqrt{(x - x_i)^2 + (y - y_i)^2}, i = 1, 2, ..., n \tag{3.7}$$

Combining Equations (3.6) and (3.7), we have the following unsolved equation.

$$\sqrt{(x - x_i)^2 + (y - y_i)^2} + ct_s - ct_i + \xi = 0 \tag{3.8}$$

To calculate a 2D position, we require three access points to transmit signals. Similarly, instead of using incident angles, we use triangulation approach with the time of arrival to estimate the user position. Figure 3.5 shows 2D positioning case using three access points (APs). The position of these three APs' location are denoted as (x_1, y_1), (x_2, y_2), (x_3, y_3) and the user position is (x, y).

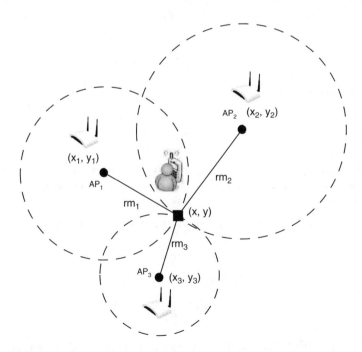

Figure 3.5   Determining 2D position of a user using TOA method.

In order to solve three unknown values (x, y, ξ), we need to solve following equations:

$$\sqrt{(x - x_1)^2 + (y - y_1)^2} + ct_s - ct_1 + \xi = 0$$
$$\sqrt{(x - x_2)^2 + (y - y_2)^2} + ct_s - ct_2 + \xi = 0 \qquad (3.9)$$
$$\sqrt{(x - x_3)^2 + (y - y_3)^2} + ct_s - ct_3 + \xi = 0$$

The unknowns (x, y, ξ) in the above nonlinear equations can be solved by three methods. They are closed-form solutions, iterative techniques based on linearization (discussed in Chapter 8, Section 8.3.4) and hyperbolic (discussed in Section 3.2.4).

There are several drawbacks to the common TOA method. First, it degrades the positioning accuracy dramatically when the positioning system is in the NLOS environments. Second, it needs time synchronization of all devices which implies having less flexibility of hardware requirement. Third, the signal arrival time to the mobile device is difficult to retrieve precisely in practice because of the variations in designs of mobile device from different manufacturers.

3.2.3 Phase of Arrival (POA)

Phase of arrival method makes use of the phase difference to estimate the position. We use the same example as shown in Figure 3.5, instead of measuring the time, we measure the phase difference, ϕ_i between AP_i and the target object.

Let $s_i(t)$ be the received signal from ith AP to the target object in a time unit of t. The received signal $x_i(t)$ emit sinusoidal signals with frequency f. The received signal $s_i(t)$ is given by:

$$s_i(t) = \sin(2\pi ft + \phi_i) \qquad (3.10)$$

The phase difference, ϕ_i is given by

$$\phi_i = \frac{2\pi f \ell_i}{c} - \xi \qquad (3.11)$$

where c is the speed of light, ℓ_i is the range measurement from ith AP to the target object, ξ denotes the error of phase measurement. We have three unknowns in this equation. ℓ_i consists of two unknown (x, y) and ξ is the unknown phase measurement error.

Equation (3.12) is used to form the following equations to estimate the 2D position.

$$\begin{aligned}
\ell_1 &= \frac{c(\phi_1 + \xi)}{2\pi f} \\
\ell_2 &= \frac{c(\phi_2 + \xi)}{2\pi f} \\
\ell_3 &= \frac{c(\phi_3 + \xi)}{2\pi f}
\end{aligned} \qquad (3.12)$$

Similar to TOA method, POA method requires line-of-sight (LOS) path, otherwise the measurement of phase change will cause errors.

3.2.4   Time Difference of Arrival (TDOA)

Time difference of arrival (TDOA) method is a modification from TOA method. Instead of using the absolute arrival time, TDOA method estimates the position by the difference of arrival times from multiple WLAN receivers. The timing error of TOA can largely be reduced in the time difference operation. Wi-Fi sensors/tags act as receivers of WLAN signal which are usually used in the TDOA method.

There are two types of method to calculate the time difference. First, the simplest way to calculate the time difference is to directly subtract the signal arrival times from two receivers. Then the time difference will be converted into range difference. But in a real scenario, the receiver may not have timing reference of the source which makes the first way impossible to implement. The second way is to use the cross-correlation technique of two signal to find the range difference. We briefly describe this in the section below.

Cross-correlation Technique

Let $s(t)$ be the WLAN signal that transmits the received signal $x_i(t)$ at ith receiver(sensor). The received signal $x_i(t)$ experiences an interference or noise $n_i(t)$ with a time delay α_i. The received signal $x_i(t)$ is given by

$$x_i(t) = s(t - \alpha_i) + n_i(t) \tag{3.13}$$

Similarly, the received signal $x_j(t)$ at jth receiver(sensor) is given by $x_j(t) = s(t - \alpha_j) + n_j(t)$. The cross-correlation function l_{x_i,x_j} of two signals is given by integrating the lag product of $x_i(t)$ and $x_j(t)$.

$$\ell_{x_i,x_j}(\tau) = \int_{-\infty}^{\infty} x_i(t)x_j(t - \tau)dt \tag{3.14}$$

where τ is the argument to maximize l_{x_i,x_j}.

We can only observe the signal for a finite time period, T. Equation (3.14) is now transformed to be:

$$\ell_{x_i,x_j}(\tau) = \frac{1}{T} \int_0^T x_i(t)x_j(t - \tau)dt \tag{3.15}$$

The cross-correlation technique helps to find the range difference without using timing reference of source. It only requires the sensors (receivers) to share a precise time reference and reference signals. Frequency domain processing techniques are usually used to calculate τ.

Hyperbolic Equation Solving Algorithms

After we obtain the range difference by either using direct substraction of TOA or using cross-correlation technique, the constant range difference between two Wi-Fi sensors and the source forms a hyperboloid. The source can be estimated from intersections of two or more hyperbolas, assuming that the source and one Wi-Fi sensor are coplanar. This method is also called a hyperbolic position location method.

Expanding the same equation of range measurement (3.7):

$$\ell_i = \sqrt{(x_i - x)^2 + (y_i - y)^2}$$
$$= \sqrt{x_i^2 + y_i^2 - 2x_i x - 2y_i y + x^2 + y^2} \qquad (3.16)$$

where (x_i, y_i) represents the coordinate of the fixed ith Wi-Fi sensor and (x, y) denotes the coordinate of the source.

The equation of the hyperboloid $\ell_{i,j}$ (range difference) is given by:

$$\ell_{i,j} = \sqrt{(x_i - x)^2 + (y_i - y)^2} - \sqrt{(x_j - x)^2 + (y_j - y)^2} = c(t_i - t_j) \qquad (3.17)$$

where (x_j, y_j) represents the coordinate of the fixed jth Wi-Fi sensor; t_i and t_j are the time when signal receives from the ith and jth access points and c is the speed of light.

The range difference between the ith AP and the 1st AP where signal arrives first is:

$$\ell_{i,1} = \ell_i - \ell_1$$
$$= \sqrt{(x_i - x)^2 + (y_i - y)^2} - \sqrt{(x_1 - x)^2 + (y_1 - y)^2} \qquad (3.18)$$

Squaring both sides, Equation (3.18) can be re-arranged as:

$$\ell_{i,1}^2 + 2\ell_{i,1}\ell_1 + \ell_1^2 = x_i^2 + y_i^2 - 2x_i x - 2y_i y + x^2 + y^2 \qquad (3.19)$$

Subtracting Equation (3.16) when $i = 1$ from Equation (3.19):

$$\ell_{i,1}^2 + 2\ell_{i,1}\ell_1 = x_i^2 + y_i^2 - 2x_{i,1}x - 2y_{i,1}y + x^2 + y^2 \qquad (3.20)$$

Figure 3.6 shows three fixed Wi-Fi sensors, s_1, s_2, s_3 to detect the WLAN signal from a source and two pairs of hyperbolas are formed from three TDOA measurements which intersect and locate the source at (x, y). Assume we know the coordinates of three fixed Wi-Fi sensors, we can calculate the intersection point.

Using Equation (3.20), we can simplify the calculation of interaction with the following equations:

$$\ell_{2,1}^2 + 2\ell_{2,1}\ell_1 = x_2^2 + y_2^2 - 2x_{2,1}x - 2y_{2,1}y + x^2 + y^2$$
$$\ell_{3,1}^2 + 2\ell_{3,1}\ell_1 = x_3^2 + y_3^2 - 2x_{3,1}x - 2y_{3,1}y + x^2 + y^2 \qquad (3.21)$$

where $x_{i,1} = x_i - x_1$ and $y_{i,1} = y_i - y_1$

The set of equations (3.21) are now linear. We only need to solve two unknowns, they are the source location (x, y) and the range measurement ℓ_1 from the 1st sensor to the source.

The unknowns (x, y, ℓ_1) in the above linear equations can be solved by two methods. They are Fang's Method (Fang et al., 1990) and Chan's Method (Chan and Ho, 1994). Chan's method is better than Fang's method because it can use redundant measurements to simplify the calculation. Also, it provides a closed-form solution and is less computational-intensive to compute both close and distant position of the source.

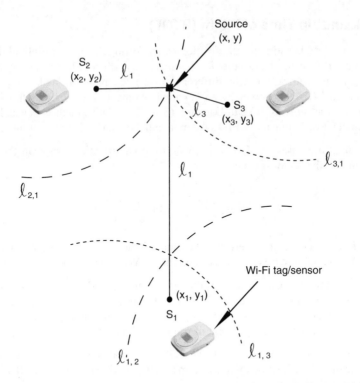

Figure 3.6   Determining 2D position of a source using TDOA method.

Chan's method begins with following solution that is extended from Equation (3.21):

$$\begin{bmatrix} x \\ y \end{bmatrix} = -\begin{bmatrix} x_{2,1} & y_{2,1} \\ x_{3,1} & y_{3,1} \end{bmatrix}^{-1} \times \left\{ \begin{bmatrix} \ell_{2,1} \\ \ell_{3,1} \end{bmatrix} \ell_1 + \frac{1}{2} \begin{bmatrix} \ell_{2,1}^2 + x_1^2 + y_1^2 - x_2^2 - y_2^2 \\ \ell_{2,1}^2 + x_1^2 + y_1^2 - x_3^2 - y_3^2 \end{bmatrix} \right\} \quad (3.22)$$

We substitute the value of (x, y) in the term of ℓ_1 into Equation 3.16 at $i = 1$. A quadratic equation in term of ℓ_1 is produced. A quadratic equation will be in this form: $a\ell_1^2 + b\ell_1 + c = 0$. The equation may have two positive roots for ℓ_1 which is

$$\ell_1 = \frac{-b \pm \sqrt{b^2 - 4ac}}{2a} \quad (3.23)$$

We use the initial condition to determine which root satisfies the priori information. After we find the right root, we substitute back into Equation (3.22) to calculate (x, y).

We also provide an alternative method, Linearization in Chapter 8, Section 8.3.4 to solve the solution.

3.2.5 Roundtrip Time of Flight (RTOF)

Roundtrip time of flight estimates the position by measuring the time of flight of the WLAN signal from a AP (transmitter) to the WLAN receiver and back. It can largely reduce the timing error by measuring the time difference of roundtrip. It can make sure the signal is propagated to WLAN receiver by an acknowledgement signal or an echo. The range measurement is the same as TOA in Equation (3.7). The TDOA method is usually applied in radar, vehicle positioning and RFID system to find the relative position of mobile objects.

Using the range measurement Equation (3.7), the roundtrip time between the unknown Wi-Fi sensor and the ith Wi-Fi access point is:

$$t_{roundtrip_i} = \frac{\ell_{i,0} + \ell_{0,i}}{c} + \xi = \frac{2\ell_i}{c} + \xi \tag{3.24}$$

where ξ is the time measurement error of the roundtrip and $\ell_{i,0} = \ell_{0,i}$ is the range measurement between the unknown Wi-Fi sensor and the ith Wi-Fi access point.

Figure 3.7 shows the positioning system that determines the 2D position of a Wi-Fi sensor using the RTOF method. Similarly, RTOF can be used together with the triangulation approach to estimate the position. The same set of triangulation algorithms will be set as Equation (3.9) but with the replacement of roundtrip time.

Similar to the TOA and TDOA methods, the RTOF method degrades the positioning accuracy dramatically when the positioning system is in the NLOS environments. It also requires time synchronization of all devices.

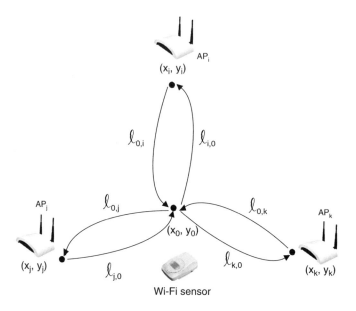

Figure 3.7   Determining 2D position of a Wi-Fi sensor using RTOF method.

3.3   Location-fingerprinting-based Algorithms

Assume a Wi-Fi enabled device always receives the same signal strength in the same location (i.e., the RSSs and coordinates), such RSSs and coordinates would be served as a unique 'fingerprint' of this location. We can collect a 'fingerprint' in each location and store them in a database. Every time we come to a new location, we detect the Wi-Fi signal and estimate the location by measuring the similarity between current and stored fingerprints.

In other words, location fingerprinting (Kaemarungsi and Krishnamurthy, 2004; Taheri *et al.*, 2004; Li *et al.*, 2005; Fang *et al.*, 2008; Kjaergaard and Munk, 2008; Swangmuang and Krishnamurthy, 2008) refers to techniques that allows a Wi-Fi enabled device to locate itself by accessing a database containing the fingerprints (i.e., the RSSs and coordinates) and then calculating its own coordinates by comparing its LF with the LF database.

There are two phases of LF-based techniques: offline phases and online phases. The first offline phase detects IEEE 802.11b Wi-Fi signal strength and collect the location fingerprints for a training database. In the second online phase, the location fingerprints are retrieved by the mobile Wi-Fi enabled device and the location is estimated by matching similarity to the LF training database.

There are at least five LF-based positioning methods, the K-Nearest Neighbor (K-NN), the probabilistic method, neural networks, support vector machine (SVM) and the smallest M-vertex polygon (SMP). Figure 3.8 shows five different types of location fingerprinting method.

3.3.1   K-Nearest Neighbor Algorithms

The K-Nearest Neighbor algorithm requires the online RSS to search for K closest matches of known locations in signal space from the previously-built database according to root mean square errors principle. By averaging these K location candidates with or without adopting the distances in signal space as weights, an estimated location is obtained via weighted K-NN or unweighted K-NN. K-NN classifies a new data point based on the majority of its K-nearest neighbors. For different applications, different distance functions are defined to quantify the 'similarity' between the training and testing points (Lin *et al.*, 2009).

In the simplest case ($K = 1$), the algorithm finds the single closest match and uses that fingerprint's location as prediction. In this algorithm, K is the parameters adapted for better performance.

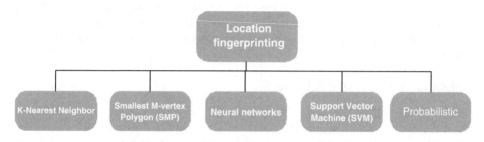

Figure 3.8   Five types of location fingerprinting method.

More specifically, the first set of data is the offline samples of RSS from N APs in the area. Each element in a vector is an independent RSS (in dBm) collected from APs in the location. $S = \{s_1 ... s_n | s_i \in \Re^n\}$ is a set of online sampling LF vectors in database. The second set of data contains online RSSs, $R = \{r_1 ... r_n | s_i \in \Re^n\}$ from n APs at a particular location.

The K-NN algorithm requires the collection of data $\{(s_i, v_q), i, q \in N\}$, for n locations in the site, where v_q is the known location of the q'th measurement and the vector s_i is the 'fingerprint' of the location v_i. When a receiver in unknown location A becomes aware of a new fingerprint r, it searches for the fingerprint s_i that is closest to r and then estimates the location. The unknown location for r is decided by a majority vote from the K shortest distance fingerprints.

We can estimate the location v_q by clustering the distance between online received LF vector r and offline sampling LF vector s_i as

$$v_q (r, f) = \left(\sum_{i=1}^{n} |r - s_i|^q \right)^{\frac{1}{q}} \tag{3.25}$$

v_q is called Manhattan distance if $q = 1$ and Euclidean distance if $q = 2$; the accuracy does not necessarily become higher as q increases.

Let s_{ij} be jth sample RSS in the ith access points. m is the number of access points. n is the number of sample data. The distance between s_i and s_{ij} is defined as

$$v_j = \sqrt{\sum_{i=1}^{m} s_{ij} - s_i} \quad j = 1 ... n \tag{3.26}$$

Electing K samples from the smallest value and calculate average coordinates as outputs in following equation:

$$(x, y) = \frac{1}{K} \sum_{i=1}^{K} (x_i, y_i) \tag{3.27}$$

where (x_i, y_i) is coordinates corresponding to i th sample.

K-NN is simple to implement and it provides reasonable accuracy. However, one drawback of the standard is that K-NN suffers from signal fluctuations because the RSSs detected in the same location can be varied from time to time. The fluctuations may cause errors in location estimation. This can be partially overcome by having multiple fingerprint sets for a given location, taken at different times, assuming that one or other fingerprint may cover that fluctuation.

Redpin Algorithm - AP Similarity

The Redpin algorithm (Lin et al., 2009) is a variation of the standard K-NN algorithm where the Euclidean distance is augmented with a bonus factor to reward training and testing fingerprints to have common APs and a penalty factor for not-common APs in two fingerprints. Thus, in addition to the signal strength, the number of common access points

(NCAP) and the number of not-common access points (NNAP) also contribute to identifying the similarity of two fingerprints. The Redpin algorithm chooses $K = 1$ to decide the best match and works as follows. A mapping function $\delta(r)$ could be defined as

$$\delta(r) = \begin{cases} 0, \ s = 0 \\ 1, \ s \neq 0 \end{cases} \tag{3.28}$$

NCAP of two fingerprints, r and s, can be expressed as

$$NCAP = \sum_{i=1}^{n} \delta(r_i) \delta(s_i) \tag{3.29}$$

NNAP of r and s can be expressed as

$$NNAP = \sum_{i=1}^{n} \delta(r_i) \oplus \delta(s_i) \tag{3.30}$$

where $\oplus$ represents the exclusive disjunction. The generalized similarity value of r and s is

$$(r, s) = \alpha \sum_{i=1}^{n} \delta(r_i) \delta(s_i) - \beta \sum_{i=1}^{n} \delta(r_i) \oplus \delta(s_i) + \gamma \Lambda(r_i, s_i) \tag{3.31}$$

Λ is a heuristic function defined in the Redpin algorithm which calculates the similarity of r and s based on the signal strengths. The factors α and γ are the bonus-weights for the common APs while β is the penalty-weight for the not-common APs. The key idea behind Redpin is using NCAP and NNAP as bonus-penalty adjustments which reduces the impact of signal fluctuations.

Weighted AP Similarity

To further reduce the impact of signal fluctuations, (Lin *et al.*, 2009) suggested a weight AP similarity method. Intensity of the APs at one location is not always the same because the environmental variations cause significant Wi-Fi signal fluctuations in the same location over time, especially inside a large building with sparse APs. Intuitively, APs with higher intensity at a location v should be weighted more in determining whether a fingerprint is located at v. The correlation between APs and locations was used as the weight for each AP. Point-wise Mutual Information (PMI) was suggested to be the correlation measurement and defined as:

$$I(v|AP) = \log \frac{P(v, AP)}{P(v) P(AP)} \tag{3.32}$$

The higher the $I(v|AP)$ value, the more likely d was associated with AP. From the historical fingerprints in the database, the $I(v|AP)$ value of each location v and AP pairs could be evaluated. The PMI value to be between 0 (least correlated) and 1 (most correlated) was normalized. PMI values were applied as weighting modifiers to the bonus of each common AP and the penalty of each not common AP. The weighted similarity value of the current measured fingerprint r and a pre-recorded fingerprint s located at v is

$$D(s, r) = \alpha \sum_{i=1}^{n} \delta(s_i) \delta(r_i) I(v|AP_i) - \beta \sum_{i=1}^{n} \delta(s_i) \oplus \delta(r_i) I(v|AP_i) + \gamma \Lambda(s_i, r_i) \quad (3.33)$$

Kalman Filter (KF)

A Kalman Filter (Welch and Bishop, 2001) is applied to adjust the trajectory path. It is used to estimate the state $x \in \Re^2$ of a discrete-time tracking process that is governed by the linear stochastic difference equation. The time update equations are

$$\widehat{x_k}^- = \widehat{x_{k-1}}$$
$$P_k^- = P_{k-1} + Q \quad (3.34)$$

And our measurement update equations are

$$K_k = P_k^- (P_k^- + R)^{-1}$$
$$\widehat{x_k} = \widehat{x_k}^- + K_k(z_k - \widehat{x_k}^-)$$
$$P_k = (1 - K_k)P_k^- \quad (3.35)$$

where Q is the process noise covariance and R is measurement noise covariance. P_k^- is the a priori estimate error covariance. P_k is the a posteriori estimate error covariance. $\widehat{x_k}^-$ is the a priori state estimate at step k. $\widehat{x_k}$ is the a posteriori state estimate at step k. K_k is the Kalman blending factor.

3.3.2 Smallest M-vertex Polygon (SMP)

Similar to the KNN algorithm, Smallest M-vertex Polygon (SMP) method requires the online received signal strength to search for the smallest M-vertex polygon closest matches of known locations in signal space from the previously-built database.

Each AP will have a set of signal strengths which has the same value from the online process. A M-vertex polygon is formed by the vertex with same signal strength from each access point. The estimated position is the centroid of the smallest polygon which has the shortest perimeter.

Figure 3.9 shows a M-Vertex polygon which is formed by 10 candidates with the same signal strength from 3 APs. Suppose the polygon in Figure 3.9 has the smallest perimeter, we then average every coordinate of the vertex and estimate the position.

3.3.3 Neural Network

Neural network is a computational model which consists of an interconnected group of neurons between inputs, hidden and output layers as shown in Figure 3.10.

Similarly, during the offline stage, we input the received signal strength of each AP and the corresponding location coordinate, for example, if we have 20 APs in an area, we input signals from Wi-Fi enable device which form a LF vector with 20 dimensions in corresponding location. Usually, a multilayer perceptron (MLP) network with one hidden layer

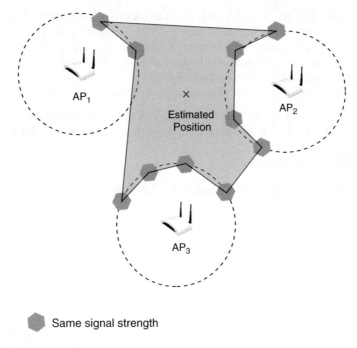

Same signal strength

Figure 3.9   M-vertex polygon with M = 3 neighbor APs.

is used for neural-networks-based positioning system. The output is the estimated location which may be a two-dimensional or three-dimensional vector. Inputs are trained with transfer function $f(x)$ and targeted for the output. The transfer function $f(x)$ is nonlinear and usually has a sigmoid shape.

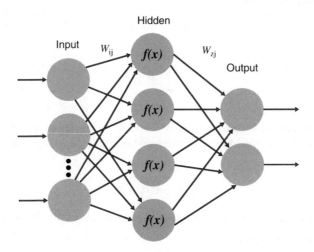

Figure 3.10   Neural network location fingerprinting method.

A back-propagation neural network (BPNN) can be used to estimate the location. A typical three-layered feed-forward neural network is shown in Figure 3.10. The nodes are the processing units, which receive input from the left side and deliver output on the right side. The input location fingerprints are applied to the input layer. The input pattern is transmitted to the input of the hidden layer through the weighted network connections. The hidden units receive the weighted pattern, where the location is estimated in an activation function.

Supposing x_j is the input node, y_i is the hidden node, o_z is the output node and t_z is the target output of ANN. The BP neural network algorithm is illustrated as below:

A) Output Node Calculation

The output nodes o_z are calculated by Equations (3.36) and (3.37).

$$y_i = f\left(\sum_{j=1}^{n} w_{ij}x_j - \upsilon_i\right) \tag{3.36}$$

$$o_z = f\left(\sum_{i=1}^{n} w_{zi}y_j - \upsilon_z\right) \tag{3.37}$$

where w_{ij} is the connection weight between node i and j, w_{zi} is the connection weight between node z and i. υ_i and υ_z are the thresholds of node i and node z respectively. f is the activation function. For simplicity, we choose sigmoid function f as an activation function. A sigmoid function is real-valued and differentiable, having either a nonnegative or nonpositive first derivative and exactly one inflection point.

$$f(t) = \frac{1}{1 + e^{-t}} \tag{3.38}$$

B) Output Node Adjustment

The adjustment of the output nodes is given by Equations (3.39), (3.40), (3.41), (3.42) and (3.43).

$$e_k = \sum_{z=1}^{n} \left|t_z^{(k)} - o_z^{(k)}\right| \tag{3.39}$$

where e_k is the training error in k iterations. $t_z^{(k)}$ and $o_z^{(k)}$ is the target and calculated output in k iterations respectively. We adjust w_{zi} to reduce the training error e_k based on the gradient descent Δw_{zj}

$$\Delta w_{zj} = -\eta\left(\partial e_k / \partial w_{zj}\right) \tag{3.40}$$

where η is the learning rate which indicates the relative size of the change in weights.

The learning rule of output nodes is:

$$\delta_z = (t_z - o_z)f'(net_z) \tag{3.41}$$

We update the next $(k + 1)$ iteration of w_{zj} and υ_{zj} as follows:

$$w_{zj}(k + 1) = w_{zj} + \Delta w_{zj} = w_{zj} + \eta \delta_z y_i \qquad (3.42)$$

$$\upsilon_{zj}(k + 1) = \upsilon_{zj} + \eta \delta_z \qquad (3.43)$$

C) Hidden Node Adjustment

The adjustment of the hidden nodes is given by Equations (3.44), (3.45) and (3.46).

$$\delta_i = y_i(1 - y_i) \sum_{z=1}^{n} \delta_z w_{zi} \qquad (3.44)$$

We update the next $(k + 1)$ iteration of w_{ij} and υ_{ij} of the hidden layer as follows:

$$w_{ij}(k + 1) = w_{ij} + \Delta w_{ij} = w_{ij} + \eta' \delta_i x_j \qquad (3.45)$$

$$\upsilon_{ij}(k + 1) = \upsilon_{ij} + \eta' \delta_i \qquad (3.46)$$

where η' is the learning rate of the hidden layers. The determination of the threshold υ_{zj} and υ_{ij} depends on the indoor positioning requirement. The network learning proceeds by training weights.

To further improve the performance of convergence, a momentum factor, β is added to Equations (3.42) and (3.45).

$$\Delta w_{zi}(k + 1) = \eta \delta_z y_i + \beta \Delta w_{zi}(k) \qquad (3.47)$$

$$\Delta w_{ij}(k + 1) = \eta' \delta_i x_j + \beta \Delta w_{ij}(k) \qquad (3.48)$$

If the resulting error is below the specified threshold υ_{zj} and υ_{ij}, then the training is stopped.

D) Location Estimation

After the training is stopped, we input the received fingerprint information to the neural network and the output yields the location. If a large location error occurs, we repeat the procedure from (A) to (C) until the network meets the threshold.

3.3.4 Support Vector Machine (SVM)

Support Vector Machine (SVM) is a machine learning methods used for classification and regressions. Intuitively, a SVM model is a representation of the examples as points in space, mapped so that the examples of the separate groups are divided by a clear gap. The idea is find a hyperplane that separates the positive examples from negative ones in the training set while maximizing the distance of them from the hyperplane. We aim to separate the location fingerprint training set and estimate the location by SVM.

Let x_i be the received signal strength (RSS) vector and n be the number of access points and can be seen as the dimension of the signal strength space, and y_i be equal to $+1$ if the location refers to a selected area, otherwise $y_i = -1$. We are given a training data set of location fingerprints, X.

$$X = \left\{(x_i, y_i) \,|\, x_i \in \mathfrak{R}^n, q_i \in \{-1, 1\}\right\}_{i=1}^n \tag{3.49}$$

$P(x_i, y_i)$ denotes the probability that $x_i \rightarrow y_i$. We define a hypothesis function $h(x, \theta)$ and the hypothesis space is $H = \{h(x, \theta), \theta \in \Theta\}$. Θ is the hypothesis index that determines whether the hypothesis holds.

Linear Separable Problems

Figure 3.11 illustrates the two-dimensional linear separable case. The norm vector $\overrightarrow{w}$ shows the direction of separating hyperplane, which is calculated by a set of support vectors. In a linear case, every separating line equation can be expressed as

$$\overrightarrow{w} \cdot x + b = 0 \tag{3.50}$$

where $\cdot$ denotes the dot product and b is the bias. As we also have to prevent data points falling into the margin, we can select the two hyperplanes of the margin in a way that there are no points between them and then try to maximize their distance. H is a classification line which separates the two types of training samples with no mistake. H_1, H_2 are the nearest straight lines to the classification line which parallels to them. The distance between H_1 and H_2 is called the gap of the two categories or classification interval (margin). The optimal separating line requires that the classification line will not only be able to separate the two categories without mistakes, but maximizes the classification interval.

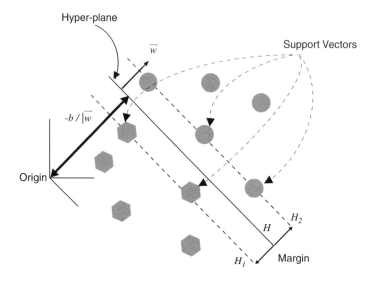

Figure 3.11   A linear SVM in a two-dimensional space.

The dot line H_1 and H_2 can be defined as

$$H_1 : \vec{w} \cdot x_i + b \geq 1 \tag{3.51}$$

or

$$H_2 : \vec{w} \cdot x_i + b \leq -1 \tag{3.52}$$

This can be further simplified as

$$y_i\left(\vec{w} \cdot x_i + b\right) \geq 1, \forall i \in [1, n] \tag{3.53}$$

The distance between these two hyperplanes (class interval) is $\frac{2}{||\vec{w}||}$. In order to maximize their distance, we need to minimize $||\vec{w}||$. However, it is quite difficult to minimize $||\vec{w}||$ because the norm vector $\vec{w}$ involves a square root. Instead of minimizing $||\vec{w}||$, we solve the following problem:

$$\begin{array}{ll} minimize & \frac{||\vec{w}||^2}{2} \\ subject\ \ to\ y_i\left(\vec{w} \cdot x_i + b\right) \geq 1, \forall i \in [1, n] \end{array} \tag{3.54}$$

Now, it becomes a quadratic programming problem. We can apply Lagrange multipliers, α and convert it into a dual form as follows:

$$\begin{array}{ll} maxmize & L(\vec{w}, b, \alpha) = \sum_{i=1}^{n} \alpha_i - \frac{1}{2} \sum_{i,j=1}^{n} \alpha_i \alpha_j y_i y_j (x_i \cdot x_j) \\ subject\ \ to & \sum_{i=1}^{n} y_i \alpha_i = 0, \alpha_i \geq 0, \forall i \in [1, n] \end{array} \tag{3.55}$$

The classification problem is now transferred to find the maximum of Equation (3.55). The (location fingerprint) vector x_i for which $\alpha_i \geq 0$ are called support vectors. If $\vec{w}^*$ denotes as the optimal value of $\vec{w}$ and α_i^* is the optimal solution to Equation (3.55), then

$$\vec{w}^* = \sum_{i=1}^{n} \alpha_i^* y_i x_i \tag{3.56}$$

The optimal linear classification function (classifier) is:

$$f(x) = sign\left((\vec{w}^* \cdot x) + b^*\right) = sign\left(\sum_{i=1}^{n} \alpha_i^* y_i (x \cdot x_i) + b^*\right) \tag{3.57}$$

Linear Inseparable Problems

If the training data set is linearly inseparable, but the linear separable hypothesis set is unchanged. In this case, we can introduce a slack variable ξ_i. Instead of solving

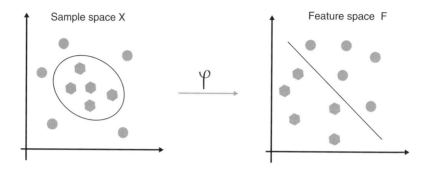

Figure 3.12   A nonlinear mapping by $\varphi(\cdot)$ function.

Equation (3.54), we solve the following problem:

$$minimize \qquad \frac{||\overrightarrow{w}||^2}{2} + C \sum_{i=1}^{n} \xi_i$$
$$subject \quad to \; y_i \left(\overrightarrow{w} \cdot x_i + b \right) \geq 1 - \xi_i, \xi_i \geq 0, \forall i \in [1, n] \tag{3.58}$$

where C is a constant to control the degree of punishment of right or wrong samples. Figure 3.12 shows the nonlinear mapping idea using $\phi(.)$ function.

Nonlinear Hypotheses

If the training dataset can only be separated by a nonlinear set, a nonlinear function, $\varphi(\cdot)$ is used to map location fingerprints from a sample space to a high-dimensional space, called the feature space. Extended to higher-dimensional space, optimal separating line will become optimal separating surface. The hyperplane formulation in a nonlinear condition becomes:

$$y_i \left(\overrightarrow{w} \cdot \varphi(x_i) + b \right) \geq 1, \forall i \in [1, n] \tag{3.59}$$

Similarly, we can apply Lagrange multipliers, α and convert it into a dual form as follows:

$$maxmize \quad L(\overrightarrow{w}, b, \xi, \alpha) = \sum_{i=1}^{n} \alpha_i - \frac{1}{2} \sum_{i,j=1}^{n} \alpha_i \alpha_j y_i y_j \varphi(x_i) \cdot \varphi(x_j)$$
$$subject \quad to \qquad \sum_{i=1}^{n} y_i \alpha_i = 0, C \geq \alpha_i \geq 0, \forall i \in [1, n] \tag{3.60}$$

The classification problem is now transferred to find the maximum of Equation (3.60).

The optimal classifier is:

$$f(x) = sign \left(\sum_{i=1}^{n} \alpha_i^* y_i \varphi(x) \cdot \varphi(x_i) + b^* \right) \tag{3.61}$$

After introducing the kernel function $k(x, y) = \varphi(x) \cdot \varphi(y)$, the classifier is:

$$f(x) = sign \left(\sum_{i=1}^{n} \alpha_i^* y_i k(x, x_i) + b^* \right) \tag{3.62}$$

The key problem of SVM application is to choose an appropriate kernel function, kernel function parameters and the penalty factor. There are four common kernel functions. They are given as follows:

(a) Dot product:$k(x, y) = x \cdot y$. No mapping is performed in this case.

(b) Radial Basis Function (RBF) kernel: $k(x, y) = e^{\left(\frac{-\|x-y\|^2}{\sigma^2}\right)}$ with variable σ.

(c) Sigmoid kernel: $k(x, y) = \tanh(ax \cdot y + b)$ with variables a and b.

(d) Polynomial kernel: $k(x, y) = (x \cdot y + 1)^{\deg}$ where the degree deg is given.

Most of SVM-based location fingerprinting chooses Radial Basis Function, as it properly maps the sample of location fingerprints into a wider space.

After we train the data set of location fingerprints, X by the above SVM algorithms, the sample of location fingerprints should be separated by SVM classifiers. Then we can make use of the trained SVM model to estimate the position.

3.3.5 Probabilistic Algorithms

Probabilistic LF calculates the most probable location out of the pre-recorded LF database. We select a location v_i if $P(v_i|s) > P(v_j|s)$, for $i, j = 1, 2, 3, ..., n, j \neq i$.

$P(v_i|s)$ denotes the probability that the WLAN-enabled device is in location d_i, given that the received signal vector is s. Also assume that $P(v_i)$ is the probability that the mobile node is in location v_i. The given decision rule is based on posteriori probability.

Using Bayes' formula (Liu et al., 2007), and assuming that $P(v_i|s) = P(v_j|s)$ for $i, j = 1, 2, 3, ..., n$ we have the following decision rule based on the likelihood that $P(s|v_i)$ is the probability that the signal vector s is received, given that the WLAN-enabled device is located in location v_i. We can estimate v_i by

$$P(v_i|S) = \frac{P(S|v_i) P(v_i)}{P(S)} \tag{3.63}$$

Since $P(S)$ is constant for all v, the algorithm can be rewritten as:

$$P(v_i|S) = P(S|v_i) P(v_i) \tag{3.64}$$

The estimated location v is the one which obtains the maximum value of the probability in Equation (3.65):

$$v = \arg \max_{v_i} [P(v_i|S)] = \arg \max_{v_i} [P(S|v_i) P(v_i)] \tag{3.65}$$

As $P(v_i)$ can be factored out from the maximization process, the estimated location v is as:

$$v = \arg\max_{v_i} [P(S|v_i)] = \arg\max_{v_i} \left[\prod_{i=1}^{n} P(s|v_i) \right] \qquad (3.66)$$

In addition to the Bayes' algorithm, kernel algorithm (Liu *et al.*, 2007) is used in calculating probability. They assume the location fingerprint database is distributed in a Gaussian distribution, therefore the mean and standard deviation of each location can be calculated. Suppose the Wi-Fi receiver measures RSS independently, we can calculate the overall probability of a location by directly multiplying the probability of all Wi-Fi receivers. Therefore, the likelihood of each location candidate can be calculated from observed signal strengths during the online stage, and the estimated location is to be decided by the previous decision rule. However, this is applicable only for discrete location candidates. Mobile units could be located at any position, not just at the discrete points. The estimated 2D location $(\hat{x}, \hat{y})$ given by (3.67) may interpolate the position coordinates and give more accurate results. It is a weighted average of the coordinates of all sampling locations:

$$(\hat{x}, \hat{y}) = \sum_{i=1}^{n} \left(P(v_i|s) \left(x_{v_i}, y_{v_i} \right) \right) \qquad (3.67)$$

3.4   Evaluation of Positioning Techniques

This section introduces the evaluation method for positioning techniques. Generally, when we evaluate whether a positioning technique is effective, we first look at its positioning accuracy. The higher the accuracy is, the more effective the system is. There are five common methods to adopt as the accuracy performance metric. They are mean square error (MSE), cumulative distribution function (CDF), Cramèr-Rao lower bound (CRLB), circular error of probable (CEP) and geometric dilution of precision (GDOP). These methods can be used to measure the accuracy performance of both indoor and outdoor positioning techniques.

3.4.1   Mean Square Error (MSE)

Mean square error (MSE) method is usually used to evaluate the performance of a positioning algorithm. Suppose we estimate a position at n times, such that we have a population of calculated coordinates, $(x_1, y_1), ..., (x_n, y_n)$. The mean of the calculated coordinates is:

$$\overline{x} = \frac{1}{n} \sum_{i=1}^{n} x_i, \overline{y} = \frac{1}{n} \sum_{i=1}^{n} y_i \qquad (3.68)$$

Let (x, y) be the actual coordinate of a position. Mean square error equation of the estimated position is given by:

$$\text{MSE}(\overline{x}, \overline{y}) = \text{E}[(\overline{x} - x)^2 + (\overline{y} - y)^2] \qquad (3.69)$$

In general expression, MSE could be rewritten as:

$$\text{MSE}(\hat{\theta}) = \text{E}[(\hat{\theta} - \theta)^2] \qquad (3.70)$$

where $\theta = [x, y]^T$, $\hat{\theta} = [\hat{x}, \hat{y}]^T$, $(\hat{x}, \hat{y})$ is the estimated coordinate. $\hat{\theta}$ is called as estimator.

The above equation gives a squared expression of error. Root mean square error (RMSE) is also used to evaluate the performance of a positioning algorithm. RMSE is a square root of the MSE:

$$\text{RMSE}(\bar{x}, \bar{y}) = \sqrt{E[(\bar{x} - x)^2 + (\bar{y} - y)^2]} \qquad (3.71)$$

3.4.2 Cumulative Distribution Function (CDF)

Cumulative distribution function (CDF) describes the probability that a real-valued random variable Z with a given probability distribution will be found at a value less than z.

$$F_Z(z) = P(Z \leq z) \qquad (3.72)$$

where Z be a numerical random variable, $F(z)$ is called the cumulative distribution function of the variable Z and $P(Z \leq z)$ denotes the probability distribution that is found at a value less than z. It can be regarded as the proportion of the population whose value is less than z. The CDF of a random variable is clearly a monotonously increasing function from 0 to 1.

We usually describe the performance of the positioning systems with how many meters of resolution with certain probability. z is the variable of error distance in meter and $P(Z \leq z)$ gives the probability of a positioning system within z meters of error distance.

$P(Z)$ can be either found by the real experiment result or by the integration of a probability density function $p(t)$ where t is the event.

$$P(Z \leq z) = \int_{-\infty}^{z} p(t)dt \qquad (3.73)$$

where t is the random variable and probability density function of a positioning system does not have the negative part, therefore the lower limit of integral is 0, not $-\infty$.

Figure 3.13 shows the relationship of cumulative distribution function (CDF) to error distance. CDF curve of Z_1 is flat increasing in error distance [4,8] and has a long tail which means the performance of positioning improves slightly and becomes steady. Z_1 reaches high probability values faster, because its distance error is concentrated in small values curve which means positioning system of Z_1 performs better than Z_2 and with less error distance. Z_1 positioning method has the precision of 85% within 4 meters.

3.4.3 Cramèr-Rao Lower Bound (CRLB)

To evaluate the accuracy of positioning algorithm, the calculated MSE or RMS is compared to the Cramèr-Rao lower bound (CRLB) to see whether a positioning algorithm is efficient to achieve a certain extent of accuracy. CRLB is to calculate the lowest possible MSE. In other words, CRLB is to calculate a lower bound on the variance of a estimator $\hat{\theta}$. However,

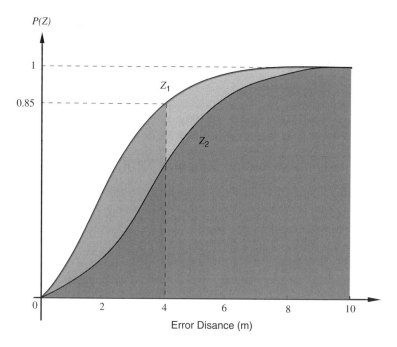

Figure 3.13   Cumulative density function of error distance.

θ consists of x and y. If the variance is required to find, we need to first find the covariance of x and y. (Variance is a special case of the covariance when the two variables are identical.)

$$
\begin{aligned}
Cov(\hat{x}, \hat{y}) &= Cov(\hat{\theta}) \\
&= \mathrm{E}[(\hat{\theta} - \theta)(\hat{\theta} - \theta)^{\mathrm{T}}] \\
&= \begin{bmatrix} \sigma_{\hat{x}}^2 & \sigma_{\hat{x}\hat{y}} \\ \sigma_{\hat{y}\hat{x}} & \sigma_{\hat{y}}^2 \end{bmatrix}
\end{aligned}
\tag{3.74}
$$

where $\sigma_{\hat{x}}$, $\sigma_{\hat{y}}$ are the standard deviation of location estimate of x and y which is usually estimated from the measurements.

The covariance of an estimator $\hat{\theta}$ is bounded by the inverse of Fisher information $I(\theta)$. The bound of inverse of Fisher information, $[I(\theta)]^{-1}$ is called CRLB.

$$
Cov(\hat{\theta}) \geq \mathrm{CRLB}(\theta) = [\mathrm{I}(\theta)]^{-1}
\tag{3.75}
$$

The Fisher information, $I(\theta)$, is the variance of the score. The score is defined as the partial derivative with respect to θ of the log of the probability density function, P_θ. The Fisher information, $I(\theta)$ is defined by:

$$
I(\theta) = \mathrm{E}\left[\left(\frac{\partial}{\partial \theta} \ln P_\theta\right)^2\right] = -\mathrm{E}\left[\frac{\partial^2}{\partial \theta^2} \ln P_\theta\right]
\tag{3.76}
$$

Because we deal with two dimensions, Fisher Information will become a $2X2$ matrix (FIM) as follows:

$$I(\theta) = \begin{bmatrix} I_{xx}(\theta) & I_{xy}(\theta) \\ I_{yx}(\theta) & I_{yy}(\theta) \end{bmatrix} \tag{3.77}$$

Different positioning method has different CRLB which depends on the definition of probability density function (PDF), P_θ.

CRLB of Time of Arrival Method

Cheung *et al.* (2004) suggests a set of PDF, FIM, CRLB for time of arrival (TOA) method.

PDF of TOA is derived from the range measurement and is defined as:

$$P_\theta(\ell) = \frac{1}{\sqrt{(2\pi)^n \prod\limits_{i=1}^{n} \sigma_i^2}} \exp\left\{ \sum_{i=1}^{n} \frac{\left[\ell_i - \sqrt{(x-x_i)^2 + (y-y_i)^2}\right]^2}{2\sigma_i^2} \right\} \tag{3.78}$$

where $P_\theta(\ell)$ is conditioned on θ where $\ell_i = [\ell_1, ..., \ell_n]^T$

The Fisher Information Matrix (FIM) for TOA is defined as:

$$I(\theta) = \begin{bmatrix} A_x & A_{xy} \\ A_{yx} & A_y \end{bmatrix} \tag{3.79}$$

where

$$A_x = \sum_{i=1}^{n} \frac{(x_i-x)^2}{\sigma_i^2[(x_i-x)^2+(y_i-y)^2]},$$

$$A_y = \sum_{i=1}^{n} \frac{(y_i-y)^2}{\sigma_i^2[(x_i-x)^2+(y_i-y)^2]},$$

$$A_{xy} = A_{yx} = \sum_{i=1}^{n} \frac{(x_i-x)(y_i-y)}{\sigma_i^2[(x_i-x)^2+(y_i-y)^2]}$$

CRLB for TOA is defined as:

$$\text{CRLB}(x) = \frac{A_y}{A_x A_y - A_{xy}^2} \tag{3.80}$$

$$\text{CRLB}(y) = \frac{A_x}{A_x A_y - A_{xy}^2} \tag{3.81}$$

CRLB of Location Fingerprinting Method

Hossain and Soh (2010) suggests a set of PDF, FIM, CRLB for location fingerprint (LF) method.

PDF of LF is derived from the propagation loss Equation (3.1) and is defined as:

$$P_\theta(r) = \prod_{i=1}^{n-1} \frac{1}{\sqrt{2\pi(\sigma_i^2 + \sigma_j^2)}} \frac{10}{\ln 10} \frac{r_j}{r_i} \exp\left\{ -\frac{\left[10\log\frac{r_i}{r_j} + 10\alpha\log\frac{d_i}{d_j} \right]^2}{2(\sigma_i^2 + \sigma_j^2)} \right\} \qquad (3.82)$$

where r_i and r_j denote the received signal strength, at the ith and jth AP which are at distance d_i and d_j.

The Fisher Information Matrix (FIM) for LF is defined as:

$$I(\theta) = \begin{bmatrix} B_x & B_{xy} \\ B_{yx} & B_y \end{bmatrix} \qquad (3.83)$$

where

$$B_x = \left(\frac{10\alpha}{(\sigma_i^2 + \sigma_j^2)\ln 10} \right)^2 \sum_{i=1}^{n-1} \left(\frac{\cos\phi_i}{d_i} - \frac{\cos\phi_j}{d_j} \right)^2,$$

$$B_y = \left(\frac{10\alpha}{(\sigma_i^2 + \sigma_j^2)\ln 10} \right)^2 \sum_{i=1}^{n-1} \left(\frac{\sin\phi_i}{d_i} - \frac{\sin\phi_j}{d_j} \right)^2,$$

$$B_{xy} = B_{yx} = \left(\frac{10\alpha}{(\sigma_i^2 + \sigma_j^2)\ln 10} \right)^2 \sum_{i=1}^{n-1} \left(\frac{\cos\phi_i}{d_i} - \frac{\cos\phi_j}{d_j} \right) \left(\frac{\sin\phi_i}{d_i} - \frac{\sin\phi_j}{d_j} \right),$$

ϕ_i and ϕ_j are the angles between the source and the ith, the jth access point respectively, $\phi_i, \phi_j \in [0, 2\pi)$. Figure 3.14 shows the angle ϕ_i between the source and the ith access point where $\phi_i \in [0, 2\pi)$.

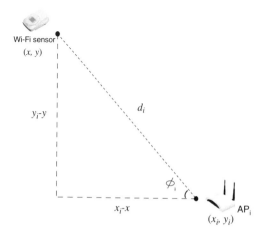

Figure 3.14   Definition of angle ϕ_i.

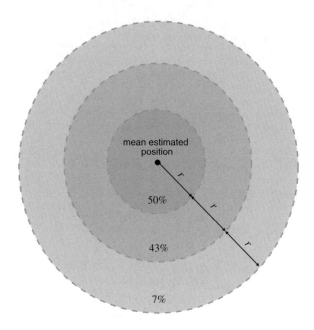

Figure 3.15 The probability of error when p(r)=50% with r meters when zero mean Gaussian random variables characterize the estimated position.

CRLB for LF is defined as:

$$\text{CRLB}(x, y) = \frac{B_x + B_y}{B_x B_y - B_{xy}{}^2} \tag{3.84}$$

3.4.4 Circular Error of Probable (CEP)

Circular error of probable (CEP) is defined as the radius of a circle which has 50% of accuracy of an estimated target around the radius of a circle, r or CEP which has its center supposed to be the true location.

In other words, the positioning system will locate in 50% of rounds within r meters of the target, 43% between r and $2r$, and 7% between $2r$ and $3r$ meters when zero mean Gaussian random variables characterize the estimated position. Figure 3.15 shows the probability of error with different radius r.

Figure 3.16 shows the geometrical relations between the actual user position, the estimated position and bias vector. If the magnitude of the bias vector is b, any estimated point must be within $b + r$. If the estimation is unbiased, $b = 0$.

Suppose zero mean Gaussian random variables characterize the estimated position. The density function of $f(x, y)$ is given by:

$$f(x, y) = \frac{1}{2\pi\sigma_x\sigma_y} \exp\left[-\frac{1}{2}\left(\frac{x^2}{\sigma_x^2} + \frac{y^2}{\sigma_y^2}\right)\right] \tag{3.85}$$

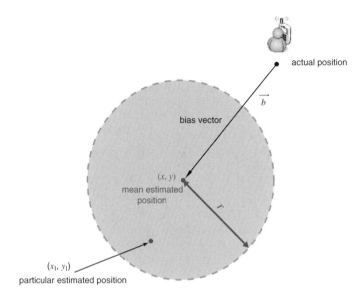

Figure 3.16 Geometry of user position, mean estimated position, CEP (r), bias vector and particular estimated position.

where $E(x) = E(y) = 0$, σ_x, σ_y are the standard deviation of location estimate of x and y which are usually estimated from the measurements.

The CEP $p(r)$ within radius is given by:

$$p(r) = \int \int_R f(x, y)dxdy \tag{3.86}$$

where R is the region within the circle.

The expression of CEP can also be rewritten as:

$$p(r) = \int \int_{x^2+y^2 \leq r^2} -\frac{1}{2\pi\sigma_x\sigma_y} \exp\left[\frac{1}{2}\left(\frac{x^2}{\sigma_x^2} + \frac{y^2}{\sigma_y^2}\right)\right] dxdy \tag{3.87}$$

Let $x = s\cos\phi$ and $y = s\sin\phi$

$$
\begin{aligned}
p(r) &= \frac{1}{2\pi\sigma_x\sigma_y} \int_0^r \int_0^{2\pi} \exp\left[\frac{1}{2}\left(\frac{s^2\cos^2\phi}{\sigma_x^2} + \frac{s^2\sin^2\phi}{\sigma_y^2}\right)\right] s\,ds\,d\phi \\
&= \frac{1}{2\pi\sigma_x\sigma_y} \int_0^r \int_0^{2\pi} s\exp\left[-\frac{s^2}{4}\left(\frac{1+\cos 2\phi}{\sigma_x^2} + \frac{1-\cos 2\phi}{\sigma_y^2}\right)\right] ds\,d\phi \\
&= \frac{1}{2\pi\sigma_x\sigma_y} \int_0^r \int_0^{2\pi} s\exp\left[-\frac{s^2}{4\sigma_x^2} - \frac{s^2}{4\sigma_y^2} - \left(\frac{s^2}{4\sigma_x^2} - \frac{s^2}{4\sigma_y^2}\right)\cos 2\phi\right] ds\,d\phi
\end{aligned}
\tag{3.88}
$$

Shnidman (1995) solves Equation (3.88) by Bessel function of zero order letting:

$$I_0(x) = \frac{1}{2\pi} \int_0^{2\pi} \exp(x \cos m\phi) d\phi \tag{3.89}$$

where $m \in N$

Substituting Equation (3.89) into Equation (3.88), we obtain

$$p(r) = \frac{1}{\sigma_x \sigma_y} \int_0^r s \exp\left(-\frac{s^2}{4\sigma_x^2} - \frac{s^2}{4\sigma_y^2}\right) I_0\left(\frac{s^2}{4\sigma_y^2} - \frac{s^2}{4\sigma_x^2}\right) ds \tag{3.90}$$

Let $t = s^2(\sigma_x^2 + \sigma_y^2)/(4\sigma_x^2\sigma_y^2)$, Equation (3.90) will be rewritten as:

$$p(r) = \frac{2\sigma_x \sigma_y}{\sigma_x^2 + \sigma_y^2} \int_0^{\frac{r^2(\sigma_x^2+\sigma_y^2)}{4\sigma_x^2\sigma_y^2}} \exp(-t) I_0\left(\frac{\sigma_x^2 - \sigma_y^2}{\sigma_x^2 + \sigma_y^2} t\right) dt \tag{3.91}$$

Shnidman (1995) simplifies Equation (3.90) by letting:

$$C_0(a, b) = \int_0^a e^{-x} I_0(bx) dx \tag{3.92}$$

and Torrieri et al. (1984) simplifies the equation with:

$$\lambda = \sigma_y/\sigma_x \tag{3.93}$$

Then, Equation (3.91) will be rewritten as:

$$p_\lambda(r) = \frac{2\lambda}{1 + \lambda^2} C_0(a, b) \tag{3.94}$$

where

$$a = \frac{r^2(1 + \lambda)}{4\lambda\sigma_x^2}, b = \frac{1 - \lambda}{1 + \lambda} \tag{3.95}$$

If we have the value of a, σ_x, σ_y and set $p(r) = 50\%$, then we can calculate the value of r which is the CEP. Since $C_0(a, b)$ is difficult to solve, Shnidman (1995) approximated $C_0(a, b)$ by the method in Pitman Jr and Trimmer (1962). When $0.3 < \lambda < 1.0$, CEP $= r = 0.615\sigma_x + 0.562\sigma_y$.

3.4.5 Geometric Dilution of Precision (GDOP)

Geometric dilution of precision (GDOP) or dilution of precision (DOP) is to evaluate the precision of a positioning system. Initially, it is used in the Global Positioning System and has been come into much wider usage in indoor positioning systems. It is defined as the ratio of the root mean square error (RMS) of position estimation in each dimension to the

Table 3.1   Meaning of GDOP values.

GDOP	Rating	Description
1	Ideal	It is an ideal situation that the positioning system always achieves highest possible precision.
1–2	Excellent	At this confidence level, position estimations are considered to be accurate enough to meet all but the most sensitive applications.
2–5	Good	This confidence level can provide an approximate positioning and is still reliable for navigation applications.
5–10	Moderate	This confidence level can only provide less accurate and reliable positioning. It can only provide a rough estimate of the current location.
10–20	Fair	At this low confidence level, positioning system can only locate the target in a very rough estimation and it is very unreliable.
>20	Poor	At this extreme low confidence level, the performance of positioning system is unacceptable and is not able to locate the target within hundreds of meters.

RMSE of range measurement. The GDOP for an unbiased estimator is given by:

$$\text{GDOP} = \frac{\sqrt{\text{E}\left[(\hat{\theta} - \text{E}(\hat{\theta}))^T(\hat{\theta} - \text{E}(\hat{\theta}))\right]}}{\sigma_{all}} \tag{3.96}$$

where $\theta = [x, y]^T$, $\hat{\theta} = [\hat{x}, \hat{y}]^T$, $(\hat{x}, \hat{y})$ is the estimated coordinate. σ_{all} is the standard deviation of range measurement.

In a 2D geometry, GDOP is given as:

$$\text{GDOP} = \frac{\sqrt{\sigma_x^2 + \sigma_y^2 + \sigma_{time}^2}}{\sigma_{all}} \tag{3.97}$$

where σ_x, σ_y are the standard deviation of location estimate of x, y. σ_{time} is the standard deviation of time measurement.

The smaller the value of GDOP is, the better performance of the positioning system is. Table 3.1 summarizes the meaning of GDOP values. We will talk further about GDOP in the Global Positioning System Chapter 7, Section 7.6.

3.5   Comparison of Indoor Positioning System

This section compares the above WLAN received signals strength positioning algorithms in terms of accuracy, precision and scalability. Accuracy is obtained by the value of mean distance errors to actual position. Precision of a positioning system is a measure of the detail and is usually obtained by many experiment trials in which the distance (in meters) is measured. We define the scope of scalability as space dimensionality of a positioning system. Having defined the common comparison principles, Table 3.2 summarizes and compares each performance of indoor positioning system.

Table 3.2   Comparison of indoor positioning system.

Propagation type

Positioning algorithm	Accuracy	Precision	Scalability
AOA (Zhang and Wong, 2009)	5 m	90% within 4.2 m	2D
TOA (Mogi and Ohtsuki, 2008)	4 m	90% within 4.5 m	2D
POA	3 m	90% within 5.9 m	2D
TDOA (Yamasaki et al., 2005)	3 m	67% within 2.4m	2D
RTOF	3 m	90% within 3.5m	2D

Location fingerprinting type

Positioning algorithm	Accuracy	Precision	Scalability
KNN (Chan et al., 2009b)	4 m	90% within 3.5 m	2D
SMP (Gwon and Jain, 2004)	2.7 m	50% within 2.7 m	2D
NN (Fang and Lin, 2008)	3 m	63% within 2 m	2D
SVM (Qiu and Kennedy, 2007)	3 m	90% within 5.12 m	2D & 3D
Probabilistic (Chan et al., 2009a)	2.8 m	90% within 2.8 m	2D & 3D

Chapter Summary

This chapter has covered the positioning algorithms of location fingerprinting and propagation based methods. We have briefly gone through the evaluation methods and comparison of each WLAN positioning system.

In the next chapter, you will learn how to use the Apple80211 framework to retrieve Wi-Fi signal strength, detect the user's orientation and implement a customized Wi-Fi positioning system in iPhone.

References

Chan, C. L., Baciu, G., and Mak, S. (2009a). Using the Newton Trust-Region Method to localize in WLAN environment. *The 5th IEEE International Conference on Wireless and Mobile Computing, Networking and Communications, WiMOB*, pp. 363–369.

Chan, C. L., Baciu, G., and Mak, S. C. (2009b). Using Wi-Fi signal strength to localize in wireless sensor networks. *The 2nd International Conference on Communications and Mobile Computing, CMC*, pp. 538–542.

Chan, Y. and Ho, K. (1994). A simple and efficient estimator for hyperbolic location. *IEEE Transactions on Signal Processing*, 42(8): 1905–1915.

Cheung, K., So, H., Ma, W., and Chan, Y. (2004). Least squares algorithms for time-of-arrival-based mobile location. *IEEE Transactions on Signal Processing*, 52(4): 1121–1128.

Fang, B. *et al.* (1990). Simple solutions for hyperbolic and related position fixes. *IEEE Transactions on Aerospace and Electronic Systems*, 26(5): 748–753.

Fang, S. and Lin, T. (2008). Indoor location system based on discriminant-adaptive neural network in IEEE 802.11 environments. *IEEE Transactions on Neural Networks*, 19(11): 1973–1978.

Fang, S., Lin, T., and Lin, P. (2008). Location fingerprinting in a decorrelated space. *IEEE Transactions on Knowledge and Data Engineering*, 20(5): 685–691.

Gwon, Y. and Jain, R. (2004). Error characteristic and calibaration-free techniques for wireless LAN-based location estimation. *2nd ACM International Symposium on Mobility Management and Wireless Access*, pp. 2–9.

Hightower, J. and Borriello, G. (2001). Location Sensing Techniques. *IEEE Computer Magazine on Location Systems for Ubiquitous Computing*, pp. 57–66.

Hossain, A. and Soh, W.-S. (2010). Cramer-Rao bound analysis of localization using signal strength difference as location fingerprint. *The 29th IEEE International Conference on Computer Communications*, pp. 1–9.

Jan, R. and Lee, Y. (2003). An indoor geolocation system for wireless LANs. *International Conference on Parallel Processing Workshops*, pp. 29–34.

Kaemarungsi, K. and Krishnamurthy, P. (2004). Modeling of indoor positioning systems based on location fingerprinting. *The 23th IEEE International Conference on Computer Communications, INFOCOM*, 2, pp. 1012–1022.

Kjaergaard, M. B. and Munk, C. V. (2008). Hyperbolic location fingerprinting: a calibration-free solution for handling differences in signal strength. *The 6th IEEE International Conference on Pervasive Computing and Communications*, pp. 110–116.

Li, B., Wang, Y., Lee, H., Dempster, A., and Rizos, C. (2005). Method for yielding a database of location fingerprints in WLAN. *IEE Proceedings on Communications*, 152(5): 580–586.

Lin, H., Zhang, Y., Griss, M., and Landa, I. (2009). WASP: an enhanced indoor locationing algorithm for a congested Wi-Fi environment. *Mobile Entity Localization and Tracking in GPS-less Environnments*, pp. 183–196.

Liu, H., Darabi, H., Banerjee, P., and Liu, J. (2007). Survey of wireless indoor positioning techniques and systems. *IEEE Transactions on Systems, Man, and Cybernetics, Part C: Applications and Reviews*, 37(6): 1067–1080.

Metsala, M. (2004). *Positioning Framework for Symbian OS*. Masters thesis, University of Turku.

Mogi, T. and Ohtsuki, T. (2008). TOA localization using RSS weight with path loss exponents estimation in NLOS environments. *The 14th Asia-Pacific Conference on Communications, APCC*, pp. 1–5.

Pitman Jr, G. and Trimmer, J. (1962). Inertial guidance. *American Journal of Physics*, 30: 937.

Prasithsangaree, P., Krishnamurthy, P., and Chrysanthis, P. (2002). On indoor position location with wireless LANs. *The 13th IEEE International Symposium, Personal, Indoor and Mobile Radio Communications*, 2, pp. 720–724.

Qiu, L. and Kennedy, R. (2007). Radio location using pattern matching techniques in fixed wireless communication networks. *International Symposium on Communications and Information Technologies, ISCIT*, pp. 1054–1059.

Shnidman, D. (1995). Efficient computation of the circular error probability (cep) integral. *IEEE Transactions on Automatic Control*, 40(8): 1472–1474.

Swangmuang, N. and Krishnamurthy, P. (2008). Location fingerprint analyses toward efficient indoor positioning. *The 6th IEEE International Conference on Pervasive Computing and Communications*, pp. 101–109.

Taheri, A., Singh, A., and Emmanuel, A. (2004). Location fingerprinting on infrastructure 802.11 wireless local area networks (WLANs) using Locus. *The 29th IEEE International Conference on Local Computer Networks*, pp. 676–683.

Torre, A. and Rallet, A. (2005). Proximity and localization. *Regional Studies*, 39(1): 47–59.

Torrieri, D., Countermeasures, U., and Center, C. (1984). Statistical theory of passive location systems. *IEEE Transactions on Aerospace and Electronic Systems*, pp. 183–198.

van de Goor, M. (2009). *Indoor Localization in Wireless Sensor Networks*, PhD thesis, Radboud University.

Welch, G. and Bishop, G. (2001). An introduction to the Kalman filter. *ACM SIGGRAPH Course Notes*.

Yamasaki, R., Ogino, A., Tamaki, T., Uta, T., Matsuzawa, N., and Kato, T. (2005). TDOA location system for IEEE 802.11 b WLAN, pp. 2338–2343.

Zhang, V. and Wong, A. (2009). Combined AOA and TOA NLOS localization with nonlinear programming in severe multipath environments. *IEEE Conference on Wireless Communications and Networking*, pp. 2219–2224.

Chapter 4

Implementation of Wi-Fi Positioning in iPhone

*A lot of times, people don't know what they want
until you show it to them.*

Steve Jobs
1998

Chapter Contents

► Performing the site-surveying using iPhone

► Implementing the location fingerprint algorithm

► Filtering with the user's orientation

► Adjusting the trajectory using the Newton Trust Region method

Positioning technologies are widely used in many iPhone applications. iPhone employs a digital compass, Wi-Fi access and Global Positioning System. The GPS technologies used in iPhone are mature and can achieve the positioning resolution from 6 to 12 meters. GPS is the most effective in relatively open and flat outdoor environments but is much less effective in closed environments such as indoor office spaces, hilly, mountainous, or built-up areas. Wi-Fi positioning seems to be one of the best alternatives. However, many problems in Wi-Fi positioning have not yet been solved and many Wi-Fi positioning systems are still in the status of prototype. Apple has released the Core Location Framework and uses it as the main core of positioning kernel, but the Wi-Fi positioning part has not been not well covered yet. In this chapter, we will build our own customized Wi-Fi positioning system using iPhone.

Algorithms of indoor positioning systems using the existing WLAN infrastructure have been covered in Chapter 3. A recall of Chapter 3: there are two approaches to locating

Introduction to Wireless Localization: With iPhone SDK Examples, First Edition. Eddie C.L. Chan and George Baciu.
© 2012 John Wiley & Sons Singapore Pte. Ltd. Published 2012 by John Wiley & Sons Singapore Pte. Ltd.

WLAN-enabled devices being propagation-based and location-fingerprinting (LF) based. Propagation-based approaches measure the received signal strength (RSS), angle of arrival (AOA), or time difference of arrival (TDOA) of received signals and apply mathematical geometry models to determine the location of the device. The drawbacks of these approaches are highly affected by the changing of internal building infrastructure, presence of humans and interference among devices leading to unstable Wi-Fi coverage and inaccurate localization. Due to these drawbacks, an LF-based type of algorithm is selected to implement in this chapter.

Implementation of the proposed Wi-Fi positioning system involves four phases:

▶ In the first phase we detect the IEEE 802.11b Wi-Fi signal strength and collect them into a training database. This training database is treated as an offline database.

▶ The second phase uses iPhone to retrieve the 'online' signal strength to estimate the user location by comparing its coordinates with the signal strengths in the offline database.

▶ The third phase makes use of the digital compass feature in iPhone to adjust the location estimation within the user's heading direction.

▶ In the final phase, we further adjust the user's trajectory and eliminate those abnormal position estimations by applying the Newton Trust Region algorithm.

This approach is able to achieve to locate the iPhone device within 2.3 m having 90% precision. (Comparing to existing Wi-Fi positioning systems, they can only achieve the resolution 4.2 m with 90% precision.) Figure 4.1 summarizes the four phases of the proposed Wi-Fi positioning system.

In the following sections, we explain thoroughly the theories, algorithms and code used in each phase.

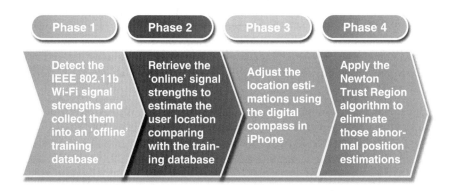

Figure 4.1   Four phases of the proposed Wi-Fi positioning system.

4.1 Site-surveying of Wi-Fi Signals Using iPhone

In this section, we talk about how to do a thorough and precise site-surveying of Wi-Fi signal strength. We can do the site-surveying by either manually or automatically (via sensors) depending on the cost and the changes of environment. If there is no any frequent changes of environments including temperature, moisture, human movements, placement of furniture and thickness of the wall, certainly, the simplest and cheapest way to do the site-surveying is the manual approach.

As mentioned in Chapter 2, the dense multi-path environment causes the received Wi-Fi signal to fluctuate dramatically; it would be very tedious and time-consuming to have an extensive training dataset surveying in every environmental changes.

The other way to do the site-surveying is via Wi-Fi sensors. We can place a set of static Wi-Fi sensors to collect the Wi-Fi signals and store them into a training database. Using the Wi-Fi sensors to retrieve the Wi-Fi signal helps to get a real-time Wi-Fi signal tracking and give a complete and accurate training database. Let's look at the configurations of Wi-Fi sensors in the market.

Wi-Fi Sensors

Every Wi-Fi sensor should operate with IEEE standard 802.11 b/g/n access and can be configured to run on the vast majority of home networks. Some of them can also even support RFID.

Many Wi-fi sensors in the market comes with a full range of security options and are able to connect with WPA2-PSK, WPA, and WEP security networks. They work in both TCP/IP and UDP connections with IPv6 and IPv4 capability. The default baud rate of the Wi-fi sensors should have 9600 baud and allowed baud rates including 9600, 19200, 38400, 57600, 115200, 230400, 460800, 921600 and should support at least 10 channels.

Each Wi-Fi sensor must have a Wi-Fi antenna to broadcast and receive the Wi-Fi signals. The Wi-Fi antenna should be installed freely from metal obstructions.

Usually, a Wi-Fi sensor has a physical power switch to turn the sensor on and off. But in some advanced ones the usage of power can be remotely adjusted. Most Wi-Fi sensors operates in ultra low power with a tiny battery from 2.4 V to 3.6V. The battery life can sustain around 2 years depending on different brands. Some even supports extra sensing features including temperature, humidity, motion, current, acceleration, and pressure, but the size of the sensor is bigger.

There are many companies providing Wi-Fi sensor products, such as Redpine, Dexter, Aginova and etc. Table 4.1 shows the specification of a Wi-Fi sensor product from Redpine.

There are two disadvantage of using Wi-Fi sensors: first is the cost and second is the power consumption. If the positioning area is huge, a lot of Wi-Fi sensors are required to install to retrieve a real-time Wi-Fi signal tracking. This may cost a lot to use more than 1000 Wi-Fi sensors to cover the entire building. Second, every sensor has a battery life and we need to replace them when it is nearly out of battery. Sometimes, replacements of the battery are tedious or even impossible because the Wi-Fi sensors may be placed in a high ceiling. No matter what kind of approach you choose, the follow-up action is to retrieve the Wi-Fi signal in the site.

Table 4.1   Specification of a Wi-Fi sensor.

Network standard support	IEEE 802.11b/g/i, draft 802.11 n/k/r
Data rates	802.11n: 6.5, 13, 19.5, 26, 39, 52, 58.5, 65 Mbps 802.11g: 6, 9, 12, 18, 24, 36, 48, 54 Mbps 802.11b: 1, 2, 5.5, 11 Mbps
Modulation techniques	OFDM with BPSK, QPSK, 16-QAM, and 64-QAM 802.11b with CCK and DSSS
QoS	WMM and WMM Power Save Support
Wireless security	802.11i: AES, TKIP, WEP, WPA, WPA2
802.11n features	MCS 0-7, STBC, RIFS,Greenfield Protection, A-MPDU, A-MSDU Aggregation with Block-ack, PSMP, MTBA
Network protocols	TCP, HTTP , ARP, UDP, IPv6
WLAN functions	Power save modes, automatic roaming, ad hoc and infras- tructure modes
Sensor interfaces	10-channel ADC, I2C, SPI, push-button, current sense
Debug interface	JTAG
Supply voltage	2.4 - 3.6 V battery
Package	LGA, 31mm x 44mm
Operating temperature	Industrial Grade $-40\,°C$ to $+85\,°C$

For the simplicity reason, we choose the manual approach which uses iPhone to retrieve the Wi-Fi signal data. The following paragraph describes about how to use the Apple80211 framework and retrieve the signal strengths, MAC addresses, Received Signal Strength Indication (RSSI) name, channels from Wi-Fi access points. If you are not familiar with iOS and Objective-C programming, you can learn more by reading Appendix A and B at the end of this book.

Using Apple80211 Framework to Retrieve the Wi-Fi Signal

Apple80211 framework provides fundamental system calls to allow developers to retrieve the real-time Wi-Fi signal in an easy way. First of all, we load the WiFiManager system library into the program so that all Apple80211 system calls are available. Then we open the framework to get an airport handle and bind it to en0 interface, and the initialization phase on framework is completed. After completing the initialization, we can retrieve the real-time Wi-Fi signal by the scan function with the airport handle, a dictionary structure object containing all the captured Wi-Fi signal information. For the dictionary structure object, it contains a set of Wi-Fi signal record per each Basic Service Set Identification (BSSID), known as MAC address, and a BSSID can uniquely identify each Basic Service Set (BSS) in the 802.11 wireless LAN. Thus, each Wi-Fi signal record is mapped with the BSSID as key in the dictionary structure. For the Wi-Fi signal record, it provides tons of information concerning this BSSID, such as Service Set Identifier (SSID), Received Signal Strength Indication (RSSI), Signal Channel, etc. RSSI is the key data for LF-based positioning system, so we will take out this data together with the BSSID and store them into the offline database. After the scanning process is completed, we can close the airport handle to release the resources.

Here is the code to use Apple80211 framework to retrieve the Wi-Fi signal:

```
#include <dlfcn.h>

// Pointers and variables for getting Wi-Fi signal
NSMutableDictionary *networks; //Key: MAC Address (BSSID)

void *libHandle;
void *airportHandle;
int (*apple80211Open)(void *);
int (*apple80211Bind)(void *, NSString *);
int (*apple80211Close)(void *);
int (*associate)(void *, NSDictionary*, NSString*);
int (*apple80211Scan)(void *, NSArray **, void *);

// Implementation of getting Wi-Fi signal
networks = [[NSMutableDictionary alloc] init];

BOOL isIOS5 = [[[UIDevice currentDevice] systemVersion] floatValue]
              >= 5.0;
if (isIOS5)
{
    // You should grant the root privilage to load the library
    // below for iOS5
    libHandle = dlopen("/System/Library/SystemConfiguration/
    IPConfiguration.bundle/IPConfiguration", RTLD_LAZY);
}
else
{
    libHandle = dlopen("/System/Library/SystemConfiguration/
    WiFiManager.bundle/WiFiManager", RTLD_LAZY);
}

char *error;
if (libHandle == NULL && (error = dlerror()) != NULL)  {
    NSLog(@"%s", error);
    exit(1);
}

apple80211Open = dlsym(libHandle, "Apple80211Open");
apple80211Bind = dlsym(libHandle, "Apple80211BindToInterface");
apple80211Close = dlsym(libHandle, "Apple80211Close");
apple80211Scan = dlsym(libHandle, "Apple80211Scan");

apple80211Open(&airportHandle);
apple80211Bind(airportHandle, @"en0");

NSLog(@"Scanning WiFi Channels...");

NSDictionary *parameters = [[NSDictionary alloc] init];
```

```
NSArray *scan_networks; // is a CFArrayRef of CFDictionaryRef(s)
        containing key/value data on each discovered network

apple80211Scan(airportHandle, &scan_networks, (__bridge void *)
        parameters);

for (int i = 0; i < (scan_networks count); i++) {
    [networks setObject:[scan_networks objectAtIndex: i] forKey:[[
    scan_networks objectAtIndex: i] objectForKey:@"BSSID"
    ]];
}

NSLog(@"Scanning WiFi Channels Finished.");

apple80211Close(airportHandle);
```

Instead of randomly taking sample points, we need to have a plan to get the signal data evenly in the site area. Engineers usually take measurements after the installation of Wi-Fi infrastructure for every 5 meters approximately. The paragraph below talks about how to grid the space effectively to achieve the target positioning resolution.

Grid the Space

Usually, we can grid the space by 10 m, 5 m, 3 m and 1 m according to the target resolution. For example, if we want to achieve the positioning precision around 3 m, we should at least have a $3x3 \, m^2$ grid to collect the Wi-Fi data. So in each grid point, we collect the coordinates, Wi-Fi signal strengths, MAC addresses, RSSI names, channels from the access points.

We have implemented an iPhone site-surveying application which allows users arbitrarily to change the grid size with zooming in and out functions. Figure 4.2 shows screen shots of a site-surveying application in iPhone.

Here is the code to zoom in and out of the griding plan:

```
// UIScrollViewDelegate for handling zoom in and out
@interface FloorPlanView : UIViewController <UIScrollViewDelegate,
        UIAlertViewDelegate, UIActionSheetDelegate> {
    UIScrollView *viewFloorPlan;
    FloorPlanGridView* gridView;
}

- (void)viewDidLoad
{
    [super viewDidLoad];
    // Do any additional setup after loading the view from its nib.

    self.title = @"Floor Plan";

    UIImageView *planView = [[UIImageView alloc] initWithImage:[[
            WiFiData getInstance]floorPlan:0].image];
```

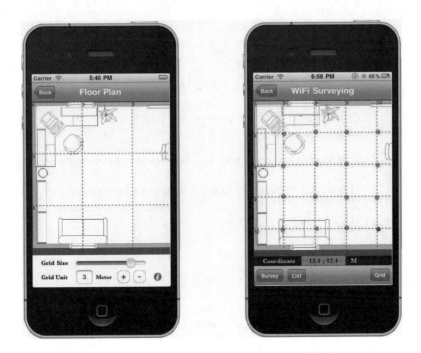

Figure 4.2   Wi-Fi Site-surveying application in iPhone.

```
[planView setTag:ZOOM_VIEW_TAG];
[planView setUserInteractionEnabled:TRUE];

// init floor plan view
[viewFloorPlan setDelegate:self];
[viewFloorPlan setContentSize:[planView frame].size];
// choose minimum scale so image width fits screen
float minScale  = [viewFloorPlan frame].size.width  / [planView
        frame].size.width;
[viewFloorPlan setMinimumZoomScale:minScale];
[viewFloorPlan addSubview:planView];
[viewFloorPlan setMaximumZoomScale:MAX_ZOOM_SCALE];

// init grid view
gridView = [[FloorPlanGridView alloc] initWithFrame:[
        viewFloorPlan frame]];
[gridView setUserInteractionEnabled:NO];
[gridView setLockGridScale:NO];
[[self view]addSubview:gridView];
gridUnitEdit.text = [NSString stringWithFormat:@"%.0f", [
        gridView gridUnit]];
gridWidthSlider.value = [gridView gridWidth];
gridWidthSlider.maximumValue = GRID_WIDTH_MAX;
gridWidthSlider.minimumValue = GRID_WIDTH_MIN;
```

(a) User A faces to the access point.   (b) User B faces behind the access point.

Figure 4.3   Users face to and behind the access point.

```
    // other init here
    ...
}

// Scroll view delegate for zooming the floor plan image
- (UIView *)viewForZoomingInScrollView:(UIScrollView *)scrollView
{
    return [self.viewFloorPlan viewWithTag:ZOOM_VIEW_TAG];
}
```

Direction of Wi-Fi Signal Retrieval

If we use the manual approach to collect the Wi-Fi signal in different facing direction, the received signal strength varies greatly. Even at the same point, it would have the average 10 dbm differences of the signal strength measurement if a user chooses an opposite facing direction to retrieve same source of access point. As human's body absorbs radio wave signals when the user obstructs the signal path that causes an extra attenuation.

To study the effect of user's orientation, we performed a simple experiment which had two users (user A and B) holding the same device at the same distance but facing towards and behind the access point. Figure 4.3 illustrates the experiment setting which user A and B face to and behind the same access point respectively.

In these settings, there was a line-of-sight (no obstacle) distance between the access point and receiver approximately 3 m. The measurements were done with two opposite facing directions. Table 4.2 shows the measurement result. When the AP was behind the user, the sample mean of received signal strength (RSS) was lower at −47.3 dBm compared to the highest RSS of −38.5 dBm when the user faced towards the access point. The results show that the RSS can be attenuated by 8.8 dBm in our settings due to the obstruction from the body. The signal fluctuated more with the standard deviation 5.89 dBm and the distribution

Table 4.2   Received signal strength measurement with different facing direction.

	Sample Mean (dBm)	Standard Deviation (dBm)	Skewness
Face towards the access point	−38.5	2.1	−0.6
Face behind the access point	−47.3	5.89	−1.34

of Wi-Fi signal strength skewed to left more when the AP was behind the user. This suggests that the user's orientation is crucial and should be included in collecting sample RSSs and computing the user location information.

iPhone employs a digital compass feature, so that we can make use of it and collect the user's direction information when we do a Wi-Fi signal site-surveying. The following paragraph describes how we use the built-in accelerometer and magnetometer in iPhone to handle the user direction.

Accelerometer in iPhone

An accelerometer is a device that measure both acceleration and gravity by sensing the amount of inertial force in a given direction. The accelerometer inside iPhone is a three-axis accelerometer, meaning that it is capable of detecting either movement or the pull of gravity in three-dimensions so it can report the roll and pitch.

Magnetometer in iPhone

A magnetometer is a device which measures the magnetic fields originating from the Earth. The measurement result can be used to calculate the precise orientation of the device. However, the magnetometer sometimes may be affected by local magnetic fields which originated from other electronic devices. As a result, the precise of estimation becomes worse.

A built-in magnetometer in iPhone allows to measure yaw of the device. Then you can combine the accelerometer and magnetometer readings to have roll, pitch and yaw measurements.

Given the pitch (θ) and roll (ϕ) rotational angles, we can transform the magnetic components to the local level plane coordinate system, and then, determine the azimuth, or compass direction (γ), as follows:

$$X_H = X \cos (\phi) + Y \sin (\theta) \sin (\phi) - Z \cos (\theta) \sin (\phi)$$
$$Y_H = Y \cos (\theta) + Z \sin (\phi) \tag{4.1}$$
$$\gamma = \arctan (Y_H / X_H)$$

where (X_H, Y_H) are the horizontal components of the Earth's magnetic field.

Here is the code to capture the direction of user; the WiFiSurveyList view controller adopts the Location Manager Delegate in order to manage the directional change of the device. A Location Manager is defined for controlling the start–stop of processing directional change, and a current heading property is defined to store the latest directional heading of the device. Initialization of Location Manager is done when the view is created, while the start and stop of Location Manager are taking place at the view appear and disappear methods respectively.

```
#import <CoreLocation/CoreLocation.h>

@interface WiFiSurveyList : UITableViewController <
        CLLocationManagerDelegate>

@property (strong, nonatomic) CLLocationManager *locationManager;
```

```objc
@property (atomic, readwrite) CLLocationDirection curHeading;

@implementation WiFiSurveyList

@synthesize locationManager;
@synthesize curHeading;

- (void)viewDidLoad
{
    [super viewDidLoad];
    // WiFiSurveyList init
    ...
    self.locationManager = [[CLLocationManager alloc] init];
    ...
    // heading service configuration
    self.locationManager.headingFilter =
            kCLLocationAccuracyBestForNavigation;
    // setup delegate callbacks
    self.locationManager.delegate = self;
}

- (void)viewWillAppear:(BOOL)animated
{
    ...
    [self.locationManager startUpdatingHeading];
}

- (void)viewWillDisappear:(BOOL)animated
{
    [self.locationManager stopUpdatingHeading];
    ...
}
```

There are two delegate methods in the Location Manager Delegate that are invoked when the Location Manager has heading data and the Location Manager encounters error conditions. When heading data is available, the first delegate method below will be invoked and the direction values are passed to this method. The direction values are measured in degrees starting at due north and continue clockwise around the compass. Thus, north is 0 degrees, east is 90 degrees, south is 180 degrees, and so on. A negative value indicates an invalid direction. When error condition is encountered, the second delegate method is invoked and the error object is passed to this method. Developers may display the error to the user that the current direction information is terminated or not accurate because of the strong magnetic interference.

```objc
// This delegate method is invoked when the location manager has
//      heading data.
- (void)locationManager:(CLLocationManager *)manager
        didUpdateHeading:(CLHeading *)heading {
    if (heading.headingAccuracy < 0)
        return;
    // Use the true heading if it is valid.
```

```
        self.curHeading = ((heading.trueHeading > 0) ? heading.
            trueHeading : heading.magneticHeading);
        NSLog(@"didUpdateHeading (x:%.1f, y:%.1f, z:%.1f, heading:%1.f)"
            , heading.x, heading.y, heading.z, self.curHeading);
}

// This delegate method is invoked when the location managed
        encounters an error condition.
- (void)locationManager:(CLLocationManager *)manager
            didFailWithError:(NSError *)error {
    if ([error code] == kCLErrorDenied) {
        [manager stopUpdatingHeading];
        UIAlertView *noCompassAlert = [[UIAlertView alloc]
                initWithTitle:@"Direction not available!" message:
                @" Uable to get the direction information."
                delegate: nil cancelButtonTitle:@"OK"
                otherButtonTitles:nil];
        [noCompassAlert show];
    } else if ([error code] == kCLErrorHeadingFailure) {
        UIAlertView *noCompassAlert = [[UIAlertView alloc]
                initWithTitle:@"Direction not accurate!" message:
                @" The direction could not be determined, most
                likely because of strong magnetic interference."
                delegate: nil cancelButtonTitle:@"OK"
                otherButtonTitles:nil];
        [noCompassAlert show];
    }
}
```

Number of Samples to be Taken

As Wi-Fi signal may fluctuate easily, we need to do multiple samplings in the grid point. In practice, we usually take 25 sample signal strengths within one minute and average them to store in the training database. The application should automatically take 30 sample signal strengths within one minute when we click on the grid point. It will exclude all the abnormal readings, select 25 sample signal strengths and average them to store in the training database.

In the following code, it is triggered when the user clicks the Scan button in Wi-Fi survey view and it will start the Wi-Fi scanning task in this event handler method. Wi-Fi is only available in a real device only, so it uses the defined constant to identify whether it is running on the iPhone simulator. When it is running on a simulator, an alert dialog is popped up. When it is running on a real device, it will scan and measure the actual Wi-Fi signal. Wi-Fi scanning is a time-consuming process and users have to wait for the scanning process to be completed; thus a progress indicator should be shown at the top of the screen during the process of Wi-Fi scanning. It will start a new thread to do the actual Wi-Fi scanning task in the application because it has to leave the main thread going back to work on UI related jobs, otherwise the progress indicator will never be visible.

```
- (IBAction)btnScanClick:(id)sender {
#if TARGET_IPHONE_SIMULATOR
```

```
    // Simulator code
    UIAlertView *alert = [[UIAlertView alloc] initWithTitle:@"Scan
                    WiFi" message:@"Only available in real
                    iPhone"delegate:self cancelButtonTitle:@"OK"
                    otherButtonTitles: nil];
    [alert show];
#else
    // start a thread to do the Wi-Fi survey so the UI will not
            hold up
    [[self.parentViewController view] addSubview: viewActivity];
    [NSThread detachNewThreadSelector:@selector(scanWiFi) toTarget:
            self withObject:nil];
#endif
}
```

In the function of Wi-Fi scanning, it is running on a new thread and the auto release pool in main thread is not covered. Thus starting another auto release pool is required to do the memory management, otherwise a memory leakage warning will be raised. Each Wi-Fi scanning survey will take several times and the result of signal strength RSSI is averaged among all the samples to reduce the noise. Each BSSID is associated with an average RSSI value, and this pair of data is stored in a list of Wi-Fi survey objects.

```
- (void)scanWiFi {
    // for new thread, need to re-issue auto release pool
    @autoreleasepool {
        WifiSurveyRecord *record = [[WifiSurveyRecord alloc] init];
        record.collectTime = [[NSDate alloc] init];
        record.direction = self.curHeading;

        NSMutableDictionary *meanSsid = [[NSMutableDictionary
                alloc] init];
        NSMutableDictionary *surveyList = [[NSMutableDictionary
                alloc] init];
        // get Wi-Fi sample for WS_NUM_CAPTURE number of times
        for (int i = 0; i < WS_NUM_CAPTURE; i++)
        {
            // scan Wi-Fi
            NSDictionary *sidBatch = [[[WiFiScanner getInstance]
                    scanNetworks] copy];
            NSLog(@"No. of BSSID found: %d", sidBatch.count);
            for (NSString *bssid in [sidBatch allKeys])
            {
                NSDictionary *sid = [sidBatch objectForKey:bssid];
                WifiSurvey *temp = [meanSsid objectForKey:bssid];
                if (temp == nil)
                {
                    temp = [[WifiSurvey alloc] init];
                    temp.count = 1;
                    temp.rssi = [[sid objectForKey:@"RSSI"]
                            doubleValue];
```

```
                        [meanSsid setObject:temp forKey:bssid];
                }
                else
                {
                    temp.count++;
                    temp.rssi += [[sid objectForKey:@"RSSI"]
                            doubleValue];
                }
                // save the raw wifi survey records
                [surveyList setObject:sid forKey:bssid];
        }
    }

    // calculate the mean RSSI of each BSSID
    for (NSString *bssid in [meanSsid allKeys])
    {
        WifiSurvey *temp = [meanSsid objectForKey:bssid];
        temp.rssi = temp.rssi / temp.count;

        NSMutableDictionary *sid = [surveyList objectForKey:
                bssid];
        NSString *meanRssi = [[NSString alloc] initWithFormat:
                @"%.1f", temp.rssi];
        [sid setObject:meanRssi forKey:@"RSSI"];

        // change back to positive num with max value 100 for
                easier comparison
        temp.rssi = WP_WIFI_MAX_RSSI + temp.rssi;
    }

    record.sidList = surveyList;
    [curCollectPoint.wifiSurveyList addObject:record];
    [curCollectPoint.wifiMeanList addObject:meanSsid];
    [self.tableView reloadData];

    [self.viewActivity removeFromSuperview];
    }
}
```

After we successfully built our offline training database, we implement the location finger-printing algorithm in iPhone.

4.2 Implementing Location Fingerprinting Algorithm in iPhone

In this section, we will implement the Location Fingerprinting algorithm in iPhone. Assume iPhone can receive the same signal strength in the same location (i.e., the RSSs and coordinates), this set of RSSs and coordinates will be served as a unique signature, 'fingerprint' of this location. It is known as 'Location Fingerprint.' We can collect 'Location Fingerprints

(LFs)' in each location and store them in an offline database. Every time we come to a new location, we detect the Wi-Fi signals and estimate the location by measuring the similarity between current and stored fingerprints.

K-Nearest Neighbor Algorithm

The K-Nearest Neighbor algorithm requires the online RSSs to search for K closest matches of known locations in the signal space from the previously-built database. In this implementation process, we take $K = 4$ which means we only select the 4 closest LFs to estimate the location.

In the K-Nearest Neighbor algorithm function, the current fingerprint will first transform to a positive signal strength format. This positive signal strength has a maximum value of 100, which indicates the strongest strength. Positive signal strength representation can do a direct numerical comparison with the stored fingerprints of each Wi-Fi survey point in the database.

```
+(CGPoint)calcKMeansPoint:(NSArray*)inCollectPoints :
        (NSDictionary*)curSid :(CollectPoint*)lastLocatePoint :
        (double)direction {
    // format the current RSSI to WifiSurvey object for calculation
    NSMutableDictionary *curRssi = [[NSMutableDictionary
            alloc] init];
    for (NSString *bssid in [curSid allKeys])
    {
        NSDictionary *sid = [curSid objectForKey:bssid];
        WifiSurvey *temp = [[WifiSurvey alloc] init];
        // change back to positive num with max value 100
        temp.rssi = WP_WIFI_MAX_RSSI + [[sid objectForKey:@"RSSI"]
                doubleValue];
        [curRssi setObject:temp forKey:bssid];
    }
    ...
}
```

After the conversion of the current fingerprint is completed, it is compared with the survey points in the training database one by one to find the nearest RSSI value on different SSIDs. The difference of the RSSI value is calculated by the Euclidean distance among all the SSIDs in the current fingerprint with the training database, and then the differences are stored into a list for ranking the survey point with the nearest signal strength.

```
        for (CollectPoint *point in inCollectPoints)
        {
            // add survey point to list and sort out the most
                    nearest location fingerprint
            double powRssiDiff = 0;
            int sidCount = 0;
            // loop all the wifi survey in each point
            for (NSDictionary *survey in point.wifiMeanList)
            {
                // compare RSSI of each BSSID
                for (NSString *bssid in curRssi)
```

```
            {
                WifiSurvey *pointSurvey = [survey objectForKey:
                        bssid];
                if (pointSurvey != nil)
                {
                    WifiSurvey *curWiFi = [curRssi objectForKey
                            :bssid];
                    powRssiDiff += pow(curWiFi.rssi -
                            pointSurvey.rssi, 2.0);
                    sidCount++;
                }
            }

            if (sidCount > 0)
            {
                WifiSurvey *rssiDiff = [[WifiSurvey alloc]
                        init];
                rssiDiff.rssi = sqrt(powRssiDiff);
                rssiDiff.count = sidCount;
                rssiDiff.point = point;
                [pointRssiDiff addObject:rssiDiff];

                powRssiDiff = 0;
                sidCount = 0;
            }
        }
    }
```

Finally, the signal strength difference list will be sorted in an ascending order and the nearest N points will be picked out to calculate the final coordinate. The final estimate of coordinate is found by averaging centers of the nearest N points coordinates.

Here is the code to implement the K-Nearest Neighbour algorithm:

```
if (pointRssiDiff.count > 0)
{
    // Sort the RSSI diff with ascending order
    [pointRssiDiff sortUsingComparator:^(id obj1, id obj2) {
        WifiSurvey *rec1 = obj1, *rec2 = obj2;
        if (rec1.rssi == rec2.rssi) return (NSComparisonResult)
                NSOrderedSame;
        return rec1.rssi > rec2.rssi ? (NSComparisonResult)
                NSOrderedDescending : (NSComparisonResult)
                NSOrderedAscending;
    }];

    // take the most nearest wifi survey points and take
        average
    CGPoint estPoint;
    int numPoints = pointRssiDiff.count > WP_KM_NUM_PT_POS ?
            WP_KM_NUM_PT_POS : pointRssiDiff.count;
    for (int i = 0; i < numPoints; i++)
```

```
    {
        WifiSurvey *rec = [pointRssiDiff objectAtIndex:i];
        estPoint.x += rec.point.pointCoord.x;
        estPoint.y += rec.point.pointCoord.y;
        NSLog(@"Nearest point #%d, coord=(%.1f,%.1f), RSSI diff
              =%.1f", i, rec.point.pointCoord.x, rec.point.
              pointCoord.y, rec.rssi);
    }
    estPoint.x /= numPoints;
    estPoint.y /= numPoints;
    return estPoint;
}
```

4.3 Orientation Filter

In this section, we determine the user's orientation using iPhone OS's framework and then apply the orientation filter to select the LFs within the direction. A valid location fingerprint sample is a sample that passes the filtering test based on the AP's one-hop and two-hop neighboring APs' positions within the compass direction γ (defined in Equation (4.1)). However, when a user moves, there may be some error α between the measured orientation and the actual compass direction. Figure 4.4 shows only the LFs within the facing direction are selected, even some other LFs are closer to the user.

In order to cover all possible directions, mathematically we can define an orientation region S as follows:

$$S = \frac{1}{2} \int_{-\alpha}^{\alpha} u^2 d\theta \tag{4.2}$$

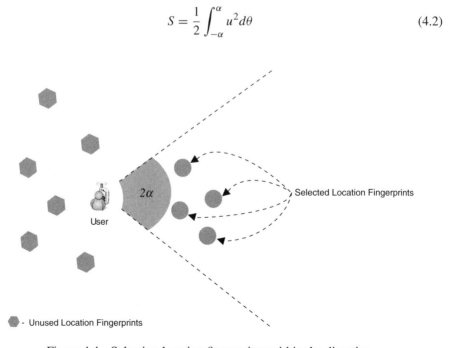

Figure 4.4   Selecting location fingerprints within the direction.

where u is the maximum distance of two-hop neighboring APs' positions. We apply the filter to this region to select the valid location fingerprint. The filter condition is as follows:

$$filter(r_i) = \begin{cases} r_i & r_i \in S, \\ 0 & r_i \notin S. \end{cases} \qquad \begin{aligned} &(4.3a) \\ &(4.3b) \end{aligned}$$

where r_i is the received LF where i belongs to any positive integer.

Here is the code to implement the orientation filter in the K-Means calculation function; each survey point in the training database is compared with the last estimated coordinates to find out the angular distance between them prior to K-Means calculation on signal strength. If the angular distance of last estimated coordinate and a survey point in the training database exceeded the boundary of current user direction, this survey point is neglected in K-Means method calculation. The remaining survey points within the boundary of user direction will be selected to do the K-Means calculation as before to obtain the final estimated coordinate. If no survey point is found in the orientation filter, all survey point will be used for the K-Means calculation. If no survey points existed in the training database, a zero coordinate will be returned which indicates an error in Wi-Fi positioning.

```
+(CGPoint)calcKMeansPoint:(NSArray*)inCollectPoints :
        (NSDictionary*)curSid :(CollectPoint*)lastLocatePoint :
        (double)direction {

    ...
    // check for orientation filter
    BOOL doOrientationFilter = lastLocatePoint != nil ? TRUE :
        FALSE;
    double curDirRadian = 0;
    CGPoint lastCoordPoint;
    if (doOrientationFilter)
    {
        curDirRadian = direction / 180.0 * M_PI;
        lastCoordPoint = lastLocatePoint.pointCoord;
    }
    ...
    // calculate the RSSI diff per each BSSID of each WiFi
            survey point
    NSMutableArray *pointRssiDiff = [[NSMutableArray alloc] init];
    BOOL searchCompleted = FALSE;
    while (!searchCompleted) {
        for (CollectPoint *point in inCollectPoints)
        {
            // filter the point by orientation filter
            if (doOrientationFilter)
            {
                // calculate the angle between current point and
                        survey point
                float theAngle = atan2(point.pointCoord.y -
                        lastCoordPoint.y,
                                        point.pointCoord.x -
                                        lastCoordPoint.x);
                // skip this point as this is not within the 120'
```

```
                                of current direction
                if (![self isWithinAngle:curDirRadian :theAngle :
                        WP_OF_HALF_ANGLE])
                    continue;
            }


            // calculate the kmeans of RSSI for this point if it is
                    selected by orientation filter
                ...
        }
    }
    ...
    NSLog(@"No reference point found for positioning...");
    return CGPointMake(0.0, 0.0);
}

+(BOOL)isWithinAngle:(double)curRadian :(double)compRadian :
        (double)angleRadian
{
    NSLog(@"Cur rad = %.2f, comp rad = %.2f, angle rad = %.2f",
            curRadian, compRadian, angleRadian);
    double angleDiff = fabs(curRadian - compRadian);
    if (angleDiff <= angleRadian || (angleDiff >= (2*M_PI -
            angleRadian) && angleDiff <= (2*M_PI + angleRadian)))
        return TRUE;
    return FALSE;
}
```

4.4   Newton Trust-Region Method

In this section, we implement the Newton Trust-Region (TR) algorithm to adjust the trajectory in a bounded region iteratively.

Current localization approaches suffer from inaccurate localization due to WLAN signal fluctuation. Signal fluctuation worsens the selection process of iterates. There are two classes of iterative algorithms, line-search and Trust Region. The TR method is more effective than other iterative nonlinear optimization methods. The TR method would be especially suitable for selecting iterates in the trajectory estimation process.

Consider a typical unconstrained minimization of location error problem,

$$\min_{x \in \vec{V}} f(x) \tag{4.4}$$

where $\vec{V}$ is a vector space. $f(x)$ is derived from the propagation-loss theorem in Chapter 2 and 3. At iteration k, with interate x_k and TR radius Δk, the TR set is:

$$A_k = \{x \in \vec{V} | \|x - x_k\|_k \leq \Delta k\} \tag{4.5}$$

There are three calculations to be made in the TR method: 1) calculation of the TR subproblem where the goal is to approximate the location minimizer in the region; 2) calculation of the TR fidelity, where the goal is to evaluate the accuracy of the location; and 3) calculation of the radius, to determine the size of a Trust Region.

4.4.1 TR Subproblem

A quadratic model m_k is constructed to approximate $f(x)$ within the TR. $<, >$ denotes the inner product. The goal of a TR subproblem is to compute whether $x_k + s_k$ is in the region.

$$m_k (x_k + s) = m_k (x_k) + \langle g_k, s \rangle + \frac{\langle s, H_x s \rangle}{2} \qquad (4.6)$$

where $m_k(x_k) = f(x_k)$, g_k is the gradient or first derivative of $f(x)$ at x_k, and H_k is the Hessian of f or second derivative of $f(x)$ at x_k. When $H_K \neq 0$, m_k is said to be a second-order model. A TR subproblem is then to compute an s_k, the TR region subproblem is:

$$s_k = \arg\min \psi_k (s) \overset{\Delta}{=} \langle g_k, s \rangle + \frac{\langle s, H_x s \rangle}{2} \qquad (4.7)$$

4.4.2 TR Fidelity

The trial point will be tested to see if it is a good candidate for the next iteration. This is evaluated by:

$$p_k = \frac{f(x_k) - f(x_k + s_k)}{m_k(x_k) - m_k(x_k + s_k)} \qquad (4.8)$$

Suppose an initial Trust Region is given and let η_1, η_2, γ_1, and γ_2 be some constants satisfying $0 < \eta_1 < \eta_2 < 1$ and $0 < \gamma_1 < \gamma_2 < 1$.

If $p_k \geq \eta_1$, then the trial point is accepted, i.e., $x_{k+1} = x_k + s_k$. Otherwise, $x_{k+1} = x_k$. When m_k approximates f well and yields a large p_k, the TR radius will be expanded for the next iteration. Otherwise, if $p_k < \eta_1$ or $p_k < 0$, m_k does not appoximate f well within the current region A_k. Therefore, the iterate remains unchanged and the TR radius will be reduced so as to allow the derivation of a more appropriate model and subproblem for the next iteration.

4.4.3 TR Radius

We can update the TR radius as follows:

$$\Delta_{k+1} = \begin{cases} \gamma_2 \Delta_k & \text{if } p_k \geq \eta_2, \\ \Delta_k & \text{if } p_k \in [\eta_1, \eta_2], \\ \gamma_1 \Delta_k & \text{if } p_k < \eta_1 \end{cases} \qquad (4.9)$$

where η_1 and η_2 represent the lower and upper bound of TR fidelity. γ_1 and γ_2 represent the changing ratio of the TR radius. The iterative process will be repeated until the sequence of iterates x_k converges. We manually choose $\eta_1 = 0.3$, $\eta_2 = 0.95$ and $\gamma_1 = 0.7$, $\gamma_2 = 1.5$ respectively, because this setting usually obtains the better result in many of our trials. There are two cases in the TR method. Figure 4.5 shows two cases, outside or within the Trust Region. A_1 is the priori estimated location, B_1 is the amended location of A_1 after TR method, A_2 is the current estimated location and B_2 is the amended location of A_2 after TR method.

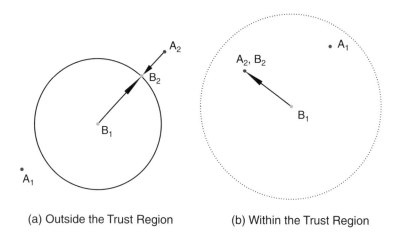

(a) Outside the Trust Region   (b) Within the Trust Region

Figure 4.5   (a) The estimated point outside the Trust Region. (b) The estimated point within the Trust Region.

First, we tested A_2 to see whether it was within the Trust Region. If it was within the Trust Region, then the current location was not required to amend as illustrated by Figure 4.5(b). Otherwise, A_2 was required to amend by TR radius which means finding the coordinate of B_2, as illustrated by Figure 4.5(a).

Second, the radius of TR method was adjusted to select the next iterates within Trust Region by (4.9). The distance ratio between $|A_1 - A_2|$ and $|B_1 - B_2|$ was calculated.

If the ratio was greater than η_2 , it means the new iterates were far away from the Trust Region; therefore, we enlarged the region by increasing the radius of Trust Region by η_2. If the ratio was smaller than η_1, it means the new iterates were almost near to the center of Trust Region; therefore, we shrank the region by decreasing the radius of Trust Region by η_1. Otherwise, the radius of Trust Region was not changed. Figure 4.6 shows three cases of the radius adjustment.

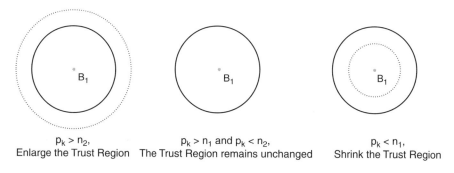

$p_k > \eta_2$, $p_k > \eta_1$ and $p_k < \eta_2$, $p_k < \eta_1$,
Enlarge the Trust Region   The Trust Region remains unchanged   Shrink the Trust Region

Figure 4.6   Three cases of radius adjustment.

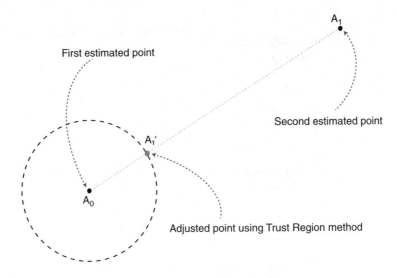

Figure 4.7   Outside the Trust Region case.

Example Cases of Appling the Trust-Region Method

This section shows the example case of how to apply the TR method to improve the trajectory. After we estimate the first point A_0, A_0 forms a trust region (defined by Equation (4.7)). Then the user moves to the other place and a second position estimation A_1 is estimated. However, A_1 is far away from A_0's trust region. (A_1 is tested by Equation (4.8) to see whether it is within the trust region.) Then A_1' is calculated and replaces A_1 to be the next estimation. Figure 4.7 shows the result of the case when A_1 is far away from the trust region. Because last time, the trust region is not able to cover the estimation, A_1' will form a bigger trust region using Equation (4.9).

When the user moves to another position, the third estimated position A_2 is calculated. This time, it is within A_1''s trust region. Therefore, we do not need to have any further adjustment. As A_2 is closed to A_1', A_2 will form a smaller trust region. Figure 4.8 shows the result of the case when A_2 is closed to A_1'.

Here is the code to implement the Trust Region method; the trust region is simplified as a circle with a certain length of radius for ease of calculation in implementation. As solving the TR subproblem is a computational intensive process, we simplify it to find whether a point is within a circle where the initial center point is the first estimated coordinate and the initial radius is 2 m.

```
+(void)calcTrustRegionPoint:(NSArray*)inLocatePoints
{
    if (inLocatePoints == nil || inLocatePoints.count < 2)
        return;

    CollectPoint *curPoint = [inLocatePoints objectAtIndex:
            inLocatePoints.count - 1];
    CollectPoint *lastPoint = [inLocatePoints objectAtIndex:
```

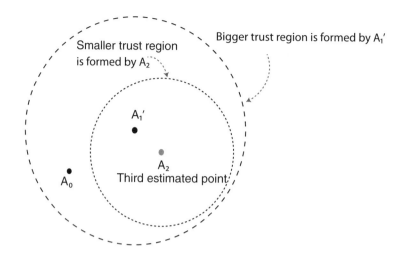

Figure 4.8   Inside the Trust Region case.

```
          inLocatePoints.count - 2];
CGPoint curGridCoord = curPoint.gridCoord;
CGPoint lastGridCoord = lastPoint.gridCoord;
CGFloat coordScale = curPoint.pointCoord.x / curPoint.
          gridCoord.x;
double distance = sqrt(pow((curGridCoord.x - lastGridCoord.x),
          2.0) + pow((curGridCoord.y - lastGridCoord.y), 2.0));
double distRatio = distance / lastPoint.TRRadius;
NSLog(@"Point distance = %.1f, ratio = %.1f", distance,
          distRatio);

if (distRatio < WP_TR_LOW_RATIO)
{
    curPoint.TRRadius = lastPoint.TRRadius * WP_TR_SHINK_RATIO;
    if (curPoint.TRRadius < WP_TR_MIN_RADIUS)
        curPoint.TRRadius = WP_TR_MIN_RADIUS;
}
else if (distRatio >= WP_TR_LOW_RATIO && distRatio <
          WP_TR_HIGH_RATIO)
{
    // no change on trust region radius
    curPoint.TRRadius = lastPoint.TRRadius;
}
else
{
    curPoint.TRRadius = lastPoint.TRRadius * WP_TR_GROW_RATIO;
    if (curPoint.TRRadius > WP_TR_MAX_RADIUS)
        curPoint.TRRadius = WP_TR_MAX_RADIUS;
    if (distRatio > WP_TR_ADJUST_RATIO)
    {
        CGPoint newGridPoint = [self findIntersectPoint:
```

```
                          lastGridCoord :curGridCoord :lastGridCoord :
                          lastPoint.TRRadius];
               CGPoint newCoordPoint = CGPointMake(newGridPoint.x *
                          coordScale, newGridPoint.y * coordScale);
               curPoint.gridCoord = newGridPoint;
               curPoint.pointCoord = newCoordPoint;
               curPoint.btnPoint.center = newCoordPoint;
           }
       }
}
```

Here is the simplified implementation of finding the intersection point of circumference of circle and straight line:

```
// find the intersection point of line and circle
+(CGPoint)findIntersectPoint:(CGPoint)p1 :(CGPoint)p2 :(CGPoint)c1
        :(CGFloat)r
{
    double a,b,c;
    double bb4ac;
    CGPoint i;

    i.x = p2.x - p1.x;
    i.y = p2.y - p1.y;
    a = i.x * i.x + i.y * i.y;
    b = 2 * (i.x * (p1.x - c1.x) + i.y * (p1.y - c1.y));
    c = c1.x * c1.x + c1.y * c1.y;
    c += p1.x * p1.x + p1.y * p1.y;
    c -= 2 * (c1.x * p1.x + c1.y * p1.y);
    c -= r * r;
    bb4ac = b * b - 4 * a * c;

    double mu1 = (-b + sqrt(bb4ac)) / (2 * a);

    CGPoint n1 = CGPointMake(p1.x + mu1 * (p2.x - p1.x), p1.y + mu1
            * (p2.y - p1.y));
    return n1;
}
```

What Does the Code Do?

Firstly, the distance ratio between distance of last and current estimated coordinates and the radius of the last trust region are evaluated. Secondly, the range of distance ratio helps to determine the change of radius of trust region on the next estimated point. Finally, the trust region radius will be decremented if the distance ratio is less than the lower bound factor, while the trust region will remain unchanged as the last one if the distance ratio is within the lower and upper bound factors. For the case of distance ratio exceeding the upper bound factor, the trust region radius of the next estimated coordinate is incremented and the coordinate of the next estimated point is tightened to the length of radius distance away from the last estimated point, which is laid on the intersection point of circumference

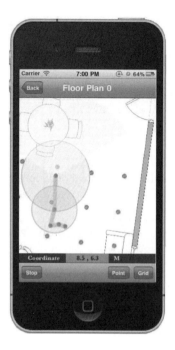 

Figure 4.9   Customized Wi-Fi positioning system in iPhone. (A full color version of this figure appears in the color plate section.)

of trust region circle of the last estimated point and the line from the last to the current estimated points. Figure 4.9 shows the customized Wi-Fi positioning system using the trust region algorithm in iPhone.

Chapter Summary

In this chapter, a customized Wi-Fi positioning system was implemented using the iPhone. You have learnt how to

- ▶ use the Apple80211 framework to retrieve Wi-Fi signal strengths,

- ▶ apply the location fingerprinting method,

- ▶ implement to detect the user's orientation, and

- ▶ adopt the Newton Trust Region method to adjust the final trajectory.

The location fingerprinting method requires the accurate signal strength measurement to achieve accurate position estimation. However, different mobile devices may vary in Wi-Fi signal measurements. The next chapter will further discuss this problem and suggest three approaches to solve it.

References

Jan, R. and Lee, Y. (2003). An indoor geolocation system for wireless LANs. *International Conference on Parallel Processing Workshops*, pp. 29–34.

Li, B., Wang, Y., Lee, H., Dempster, A., and Rizos, C. (2005). Method for yielding a database of location fingerprints in WLAN. *IEE Proceedings on Communications*, 152(5): 580–586.

Prasithsangaree, P., Krishnamurthy, P., and Chrysanthis, P. (2002). On indoor position location with wireless LANs. *The 13th IEEE International Symposium, Personal, Indoor and Mobile Radio Communications*, 2, 720–724.

Chapter 5

Positioning across Different Mobile Platform

The heaviness of being successful was replaced by the lightness of being a beginner again, less sure about everything. It freed me to enter one of the most creative periods of my life.

Steve Jobs
2005

Chapter Contents

▶ Introducing the problem of signal strength variance from different mobile platforms

▶ Calibration-free approach: signal strength value ratio

▶ Calibration-free approach: signal strength value difference

▶ Device-independent approach: Fourier descriptors

Diverse devices may receive different signal strength values even from the same source at the same condition. Chan *et al.* (2008) shows that the suggested standard deviation of received signal strength (RSS) value should be under 4 dBm for an accurate positioning. However, if the environment is in a busy traffic state, the standard deviation will be large and Kaemarungsi and Krishnamurthy (2004a,b) show that using different mobile platforms to collect RSS values may vary a 25 dB signal difference in the same location.

Most current positioning methods (Taheri *et al.*, 2004; Li *et al.*, 2005; Fang *et al.*, 2008; Kjaergaard and Munk, 2008; Swangmuang and Krishnamurthy, 2008) use the absolute received signal strength (RSS) to estimate the position coordinates and they are not accurate enough for handling the large standard deviation due to the different receiving platforms.

The location fingerprinting (LF) method locates a device by accessing a pre-recorded database containing location fingerprints (i.e., the received signal strengths and coordinates).

Introduction to Wireless Localization: With iPhone SDK Examples, First Edition. Eddie C.L. Chan and George Baciu.
© 2012 John Wiley & Sons Singapore Pte. Ltd. Published 2012 by John Wiley & Sons Singapore Pte. Ltd.

LF-based methods initially require surveying a very large training dataset and they are very sensitive to changes in the infrastructure of buildings and different receiver devices. There are two major drawbacks of LF methods. First, the absolute RSS in a time interval may not represent the IEEE 802.11 Wi-Fi signal, as the signal may fluctuate. Second, a manual error-pone calibration is needed across different mobile platforms (Dong *et al.*, 2009). The manual calibration is time-consuming and is not precise and practical to calibrate in an ad hoc manner for every new device.

There are two ways of causing signal variance, either measuring signal from access points or measuring signal by Wi-Fi receivers. Earlier research works resolved the first issue by measuring and calibrating the difference for each access point manually. There are two approaches to deal with signal variance from access points suggested by Kjaergaard and Munk (2008) and Dong *et al.* (2009). It seems that none of the research works addressed the signal variance issue from Wi-Fi receivers (nowadays, most mobile phones/devices can receive Wi-Fi). Instead of dealing with the signal variance from access points, in this chapter, we briefly introduce and cover three calibration-free solutions for handing the signal strength variance across different Wi-Fi receivers (mobile devices). They are the signal strength value ratio approach, the signal strength difference approach and our suggested approach, Fourier Descriptors (FD) (Chan *et al.*, 2010).

5.1   Signal Strength Value Ratio Approach

The most typical approach to solve the signal variance from different mobile devices is signal strength ratio calibration. As from the name, we know that we measure received signal strength (RSS) from different mobile devices to compute the signal strength ratio and calibrate the signal strength variance by the ratio.

5.1.1   Signal Strength Ratio

Suppose there are totally m Wi-Fi mobile receivers, $W = \{w_1, w_2, ..., w_m\}$ distributed in a two-dimensional model space $P = \{p_1, ..., p_n\} \in \Re^2$. P can also be denoted as a set of physical locations with (x, y) coordinates:

$$P = \{p_1 = (x_1, y_1), p_2 = (x_2, y_2), ..., p_n = (x_n, y_n)\}$$

The RSS matrix can be defined as $S = \{s_1, ..., s_m\} \in \Re^m$. s_i is the RSS matrix received by w_i. Each matrix of RSS can be denoted as $s_i = [s_{i_1}, ..., s_{i_k}] \in \Re^k$ where k is the total number of sample points. All possible signal strength values are modeled as a finite observation space, $O = \{o_1, ..., o_k\}$. An observation o in the observation space O consists of a RSS value matrix s and Wi-Fi mobile receives w. Each observation can be denoted as: $o_i = \{(w_i, s_i), ..., (w_m, s_m)\}$. The signal strength ratio, ω can be defined as follows:

$$\omega(o_a, o_b) = \frac{E(s_a)}{E(s_b)} = \frac{\frac{\sum_{i=1}^{k} s_{a_i}}{k}}{\frac{\sum_{i=1}^{k} s_{b_i}}{k}} = \frac{\sum_{i=1}^{k} s_{a_i}}{\sum_{i=1}^{k} s_{b_i}} \tag{5.1}$$

where o_a and o_b are two observations and $a < b$.

5.1.2 Log-normalized Signal Strength Ratio

The above ratio should be normalized by a log function because the signal ratios are nonlinear with respect to signal strength value which is suggested usually in Gaussian distribution (Tiemann *et al.*, 2000; Chen *et al.*, 2004; Wang *et al.*, 2006).

λ is denoted as the log-normalized signal strength ratio which is derived from Equation (5.1):

$$\lambda(o_a, o_b) = \log\left(\frac{s_{\max}}{\omega(o_a, o_b)}\right) \tag{5.2}$$

where $s_{\max}$ is the maximum value of RSS in the matrix which is used to normalize the ratio to give a positive value.

Kjaergaard and Munk (2008) make use of the above ratio to extend two LF method, probabilistic and K-Nearest Neighbor (K-NN) which are discussed in Chapter 3. Kjaergaard and Munk (2008) named it as hyperbolic location fingerprinting (HLF) approach because it is inspired from hyperbolic positioning of time difference of arrival (TDOA) method. We call the above log-normalized ratio, λ which is a 'hyperbolic' ratio.

Since the LF method has a pre-recorded database containing location fingerprints (i.e., the received signal strengths and coordinates), let la_i be the location fingerprints where

$$l_{a_i} \in L = \left\{l_{a_1}, ..., l_{a_n}\right\}$$

at ath receiver in ith location. Similarly, we can denote the location fingerprints of bth receiver in ith of location as la_i. The hyperbolic location fingerprint, h for ith location is defined as follows:

$$h_i(o_a, o_b) = \frac{1}{k} \sum_{o_a \in l_{a_i}} \sum_{o_b \in l_{b_i}} \lambda(o_a, o_b) \tag{5.3}$$

5.1.3 K-NN Hyperbolic Location Fingerprinting

K-NN location fingerprinting method usually involves finding the Euclidean distance between offline and online hyperbolic location fingerprints. We can have a new set of offline hyperbolic location fingerprints; similarly, we can obtain a new set of online hyperbolic location fingerprints from Equation (5.3). The Euclidean distance between online and offline hyperbolic location fingerprint is defined as for ith and jth location:

$$ED(o_a, o_b) = \sqrt{\sum_{o_a \times o_b} \left[h_i(o_a, o_b) - h_j(o_a, o_b)\right]^2} \tag{5.4}$$

The rest of K-NN location fingerprinting computation is quite similar to the traditional method, except that they use a new set of hyperbolic location fingerprints rather than a set of absolute value of location fingerprints. You may refer to Chapter 3, Section 3.3.1 to look at the KNN LF method.

5.1.4   Probabilistic Hyperbolic Location Fingerprinting

The probabilistic location fingerprinting approach is introduced in Chapter 3, Section 3.3.5. The algorithms are quite similar but only need to replace the offline training database by the new hyperbolic ratio location fingerprint and rebuild the histogram of observation for each Wi-Fi receiver pairs, $w_i \times w_j$ at each location p_k. The rest of calculation is the same as traditional probabilistic LF method.

5.2   Signal Strength Value Difference Approach

The second method to solve the signal variance is signal strength difference calibration approach. Dong *et al.* (2009) purely use the signal difference of two receivers and name it the DIFF approach. We use the same notation in the above and derive the equations as follows.

5.2.1   Signal Strength Value Difference

Let μ be the signal strength difference for two Wi-Fi receivers, w_a and w_b. The signal strength different g can be computed from two observations o_a and o_b as follows:

$$g(o_a, o_b) = s_{a_i} - s_{b_i}, 1 \le a < b \le m \tag{5.5}$$

The signal strength difference feature matrix G extracted from s can be given as follows:

$$G = [g(o_1, o_2), g(o_1, o_3), ..., g(o_{n-1}, o_n)] \tag{5.6}$$

5.2.2   K-NN DIFF Location Fingerprinting

The Euclidean distance between the online and offline DIFF location fingerprint is required to compute in a K-NN Location Fingerprinting approach. The Euclidean distance of DIFF between two feature matrix g_1 and g_2 is defined as follows:

$$ED(o_1, o_2) = \sqrt{\sum_{o_a, o_b, a<b} \left[g_1(o_a, o_b) - g_2(o_a, o_b) \right]^2} \tag{5.7}$$

The rest of K-NN location fingerprinting computation is quite similar as the traditional method, except using a new set of hyperbolic location fingerprint rather than a set of absolute value of location fingerprint. You may refer to Chapter 3, Section 3.3.1 to look at the KNN LF method.

5.2.3   Probabilistic DIFF Location Fingerprinting

Similar to Probabilistic Hyperbolic Location Fingerprinting, you can refer to the probabilistic location fingerprinting approach in Chapter 3, Section 3.3.5. The algorithms are quite similar but this time replace the offline training database by the new DIFF location fingerprint and rebuild the histogram of observation for each Wi-Fi receiver pairs, $w_i \times w_j$

at each location p_k. The rest of the calculation is the same as the traditional probabilistic LF method.

5.3 Fourier Descriptors Approach

The third method to solve the signal variance is the Fourier Descriptors approach which maps the absolute value of RSS into a Fourier domain. The Fourier descriptors (FD) approach uses LF-based techniques and sensors in three phases. The first phase detects the IEEE 802.11b Wi-Fi signal strength and uses a set of static LF sensors to collect the LFs into a training database in a time interval. In the second phase (location phase), the LFs (location fingerprints) will be transformed to Fourier domain and the Fourier descriptor is used to estimate the location by applying the K-Nearest Neighbor or Probabilistic LF algorithms. The Fourier descriptors describe the Wi-Fi signal.

Once we calculate the Fourier descriptors to replace the absolute value of location fingerprint, all Wi-Fi receivers will have the same Fourier location fingerprint if it is in the same condition ideally. Therefore, no further calibration is required for different receiving devices.

The Fourier descriptor is a method for measuring the shape of all or parts of an organism. It refers to the utilization of Fourier analysis, primarily the Fourier series as a curve fitting technique, that can numerically describe the shape of irregular structures such as are commonly found in living organisms. It is a well-studied algorithm for describing different wave signals and is applied variously in various fields (e.g., pattern recognition (Chen and Bui, 1999), gait recognition (Mowbray and Nixon, 2003) and image retrieval (Zhang and Lu, 2002)) but could well be applied to solve previously addressed problems in LF methods.

The Fourier descriptor is also invariant to rotation, expansion, contraction, and translation. If we keep only a low frequencies subset of descriptors, we get a curve that just approximates the outline of a shape. By increasing the number of components in the description, high frequencies are also rendered, and sharp curves or details can be generated.

5.3.1 Fourier Location Fingerprint

This subsection introduces the transformation from the absolute value of location fingerprints to Fourier location fingerprints. The Fast Fourier transform (FFT) is an algorithm that is used to attempt to determine the power versus frequency graph for a signal. For example, the FFT algorithm shows the different sounds and volumes, which are combined to make a complex sound, like the human voice. The Fourier curve is found by multiplying the periodic waveforms by the sum of the harmonically related sine waves. The Fourier transforms aims to decompose a cycle of an arbitrary waveform into its sine components. The Inverse Fourier transform converts a series of sine components into the resulting waveform. We use the FFT algorithm to convert the vector of Wi-Fi signals' amplitude of the access points and use the Fourier descriptor to recognize the pattern.

The modeling of Wi-Fi signals by the Fourier descriptors uses a pre-recorded location fingerprint database. The Wi-Fi signal in a location can be represented by a complex function of v defined by:

$$v(k) = x(k) + jy(k) \tag{5.8}$$

Table 5.1   The influence of Fourier operations.

	Signal $u(n)$	Fourier Descriptors $c(n)$
Translation	$u(n) + u(0)$	$c(0) + u(0)c(n)$
Homothetic transformation	$\lambda u(n)$	$\lambda c(n)$
Reverse description	$u(N - n)$	$e^{-j2\pi n}c(-n)$
Rotation	$e^{j\varphi}u(n)$	$e^{j\varphi}c(n)$

where $k = 0, 1..., N - 1$, N is the total number of period, x and y are the in-phase and quadratic components of the measurement Wi-Fi signal. v can be transformed into the frequency domain by the Discrete Fourier Transformation (DFT) of $c\,(n)$.

$$c\,(n) = \sum_{k=0}^{N-1} u\,(k)\,e^{j2\pi nk/2}, n = 0, 1, ..., N - 1 \tag{5.9}$$

By Euler's identity,

$$c\,(n) = \sum_{k=0}^{N-1} u\,(k)\,(\cos{(2\pi nk/N)} - j\sin{(2\pi nk/N)}) \tag{5.10}$$

The result can be transformed back into the spatial domain via the Inverse Discrete Fourier Transformation (IDFT) of $u\,(n)$.

$$u\,(n) = \frac{1}{N}\sum_{k=0}^{N-1} c\,(k)\,e^{-j2\pi nk/N}, n = 0, 1, ..., N - 1 \tag{5.11}$$

The complex coefficients $c\,(n)$ are called the Fourier descriptors (FDs) of the corresponding signal. The coefficients with low index contain information on the general form of the signal and the ones with high index contain information on the finer details of the signal. However, the first coefficient $c\,(0)$ depends only on distance from the AP of a Wi-Fi signal. Rotation invariance is obtained by ignoring the phase information. Scale invariance is obtained by dividing the magnitude values of all coefficients by the magnitude of the second coefficient $c\,(1)$. The new set of coefficients, f is given by the following equation:

$$f_k = \frac{|c\,(n)|}{|c\,(1)|}, n = 2, 3, ..., N - 1, k = 0, 1, ..., N - 3 \tag{5.12}$$

These are sensitive to transformations of the signal such as translation, homothetic transformation and reverse description. Table 5.1 summarizes the influence of these operations.

In ideal terms, all Wi-Fi receivers will have the same Fourier location fingerprint set. It is important to have enough samples and to distill key frames of the Wi-Fi signal. After samples are collected in a location, the difference between the samples and the mean is obtained and then the difference is transformed into the frequency domain. Using Equation (5.9), the value is expressed in term of complex number. The DFT points are then used to generate Fourier descriptors using Equation (5.12).

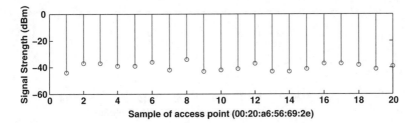

Figure 5.1   Signal samples of Wi-Fi RSS

5.3.2   Example of Fourier Location Fingerprint

We use the same notation in the above. Each observations o_i collects the signal strength difference by a Wi-Fi receiver, w_i. For example, we sample the signal strength values into a 20 x 1-point samples matrix C:

$$[-56, \; -56, \; -56, \; -56, \; -57, \; -54, \; -55, \; -54, \; 55, \; -53, \; -52, \; -58, \; -52, \; -51,$$
$$-52, \; -50, \; -51, \; -51, \; -54, \; -52]$$

Figure 5.1 shows signal samples with 20 samples in a location. C subtracts with the mean of sample. The new matrix C will be as follows:

$$[-2.25, \; -2.25, \; -2.25, \; -2.25, \; -3.25, \; -0.25, \; -1.25, \; -0.25, \; -1.25, \; -0.75, \; 1.75,$$
$$-4.25, \; 1.75, \; 2.75, \; 1.75, \; 3.75, \; 2.75, \; 2.75, \; -0.25, \; 1.75]$$

C will be transformed into $c(n)$ by Equation (5.9). Figure 5.2 shows the absolute value of each DFT point matching the original spectrum at the same location. It is the DFT magnitude spectrum of the signal. After the Fourier transform of the signal is obtained, it is converted into Fourier descriptors using Equation (5.12). The normalized Fourier descriptor describes the signal in Figure 5.3.

5.3.3   K-NN Fourier Location Fingerprinting

In theoretical terms, all receivers will have the same Fourier location fingerprint set and no adjustment is required. The rest of K-NN location fingerprinting computation is quite similar to the traditional method, except using a new set of Fourier location fingerprint

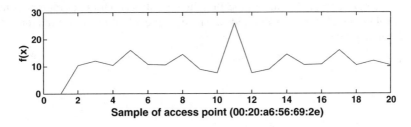

Figure 5.2   Discrete Fourier Transform (DFT) of Wi-Fi RSS

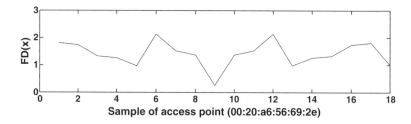

Figure 5.3   Fourier descriptors (FD) of Wi-Fi RSS

rather than a set of absolute value of location fingerprints. You may refer to Chapter 3, Section 3.3.1 to look at the K-NN LF algorithms.

5.3.4   Probabilistic Fourier Location Fingerprinting

In theoretical terms, all receivers will have the same Fourier location fingerprint set and no adjustments are required. You may refer to algorithms of the probabilistic location fingerprinting approach in Chapter 3, Section 3.3.5. The algorithms are quite similar but this time replace the offline training database by the new Fourier location fingerprint and rebuild the histogram of observation for each Wi-Fi receiver pairs, $w_i \times w_j$ at each location p_k. The rest of calculation is the same as traditional probabilistic LF method.

Chapter Summary

In this chapter, we have presented three calibration-free localization approaches for handling signal strength variance between diverse receiving devices, they are: (1) Signal Strength Value Ratio (Hyperbolic Location Fingerprinting), (2) Signal Strength Value Difference (DIFF) and finally (3) Fourier Descriptors (FD).

The ideas of each approach are listed below:

▶ Signal Strength Value Ratio: To record location fingerprints as signal-strength ratios between pairs of Wi-Fi receivers instead of absolute signal-strength values

▶ Signal Strength Value Difference: To record location fingerprints as signal strength differences between pairs of Wi-Fi receivers instead of absolute signal strength values.

▶ Fourier Descriptors: To record location fingerprints as Fourier descriptors instead of absolute signal strength values, ideally no further calibration is required for different receiving device.

References

Chan, C. L., Baciu, G., and Mak, S. (2008). Wireless tracking analysis in location fingerprint. *The 4th IEEE Wireless and Mobile Computing, Networking and Communications*, pp. 214–220.

Chan, C. L., Baciu, G., and Mak, S. C. (2010). Wi-Fi positioning based on Fourier descriptors. *The 3rd International Conference on Communications and Mobile Computing, CMC*, 3: 545–551.

Chen, G. and Bui, T. (1999). Invariant Fourier-wavelet descriptor for pattern recognition. *Pattern Recognition*, 32: 1083–1088.

Chen, W., Hou, J., and Sha, L. (2004). Dynamic Clustering for Acoustic Target Tracking in Wireless Sensor Networks. *IEEE Transactions on Mobile Computing*, pp. 258–271.

Dong, F., Chen, Y., Liu, J., Ning, Q., and Piao, S. (2009). A calibration-free localization solution for handling signal strength variance. *The 2nd International Conference on Mobile Entity Localization and Tracking in GPS-less Environments*, pp. 79–90.

Fang, S., Lin, T., and Lin, P. (2008). Location fingerprinting in a decorrelated space. *IEEE Transactions on Knowledge and Data Engineering*, 20(5): 685–691.

Kaemarungsi, K. and Krishnamurthy, P. (2004a). Modeling of indoor positioning systems based on location fingerprinting. *The 23th IEEE International Conference on Computer Communications, INFOCOM*, 2, 1012–1022.

Kaemarungsi, K. and Krishnamurthy, P. (2004b). Properties of indoor received signal strength for WLAN location fingerprinting. *The 1st International Conference on Mobile and Ubiquitous Systems: Networking and Services, MOBIQUITOUS*, pp. 14–23.

Kjaergaard, M. B. and Munk, C. V. (2008). Hyperbolic location fingerprinting: a calibration-free solution for handling differences in signal strength. *The 6th IEEE International Conference on Pervasive Computing and Communications*, pp. 110–116.

Li, B., Wang, Y., Lee, H., Dempster, A., and Rizos, C. (2005). Method for yielding a database of location fingerprints in WLAN. *IEE Proceedings on Communications*, 152(5): 580–586.

Mowbray, S. and Nixon, M. (2003). Automatic gait recognition via Fourier descriptors of deformable objects. *Lecture Notes in Computer Science*, pp. 566–573.

Swangmuang, N. and Krishnamurthy, P. (2008). Location fingerprint analyses toward efficient indoor positioning. *The 6th IEEE International Conference on Pervasive Computing and Communications*, pp. 101–109.

Taheri, A., Singh, A., and Emmanuel, A. (2004). Location fingerprinting on infrastructure 802.11 wireless local area networks (WLANs) using Locus. *The 29th IEEE International Conference on Local Computer Networks*, pp. 676–683.

Tiemann, C., Martin, S., Joseph, J., and Mobley, R. (2000). Aerial and acoustic marine mammal detection and localization on Navy ranges. *U.S. Department of Commerce and Department of the Navy, Joint Interim Report: Bahamas Marine Mammal Stranding Event.*

Wang, J., Zha, H., and Cipolla, R. (2006). Coarse-to-fine vision-based localization by indexing scale-invariant features. *IEEE Transactions on Systems, Man, and Cybernetics - Part B: Cybernetics*, 36(2): 413.

Zhang, D. and Lu, G. (2002). Generic Fourier descriptors for shape-based image retrieval. *IEEE International Conference on Multimedia and Expo*, 2: 425–428.

Chapter 6

Wi-Fi Signal Visualization

We made the buttons on the screen look so good you'll want to lick them.

Steve Jobs
Jan. 2000

Chapter Contents

▶ Why do we need a Wi-Fi visualization tool?

▶ Introducing the fuzzy color map

▶ Introducing the topographic map

▶ Refining the positioning based on the visualization result

Visualization of wireless signal strength is crucial to the post-analysis of indoor positioning systems. The visualization tool should have spatial elements to visualize the received signal strength (RSS) distribution. As a management information tool, along with the supporting log files, a displayed map should be embedded to provide an at-a-glance view of the wireless usage in different buildings, which can identify areas of oversubscription for the wireless APs in use.

6.1 Why Do We Need a Wi-Fi Visualization Tool?

With the visualization tool, engineers can evaluate the performance of indoor positioning systems. For example, the tool can easily visualize where there might be too many access points, packed too closely together in a region which might lead to overlaps of signals, cause interference and potential security risks. Some access points can even be removed to achieve the same signal stability level. On the other hand, there might be fewer access points in a region which leads to weak signal coverage. The tool can inform the installation of more wireless access points (APs) or the upgrade to higher density 802.11n APs.

Introduction to Wireless Localization: With iPhone SDK Examples, First Edition. Eddie C.L. Chan and George Baciu.
© 2012 John Wiley & Sons Singapore Pte. Ltd. Published 2012 by John Wiley & Sons Singapore Pte. Ltd.

Yet, there is currently a lack of an analytical model that addresses effectively the analysis of how the WLAN infrastructure affects the accuracy of tracking.

The fuzzy color mapping has been widely applied in other fields, such as temperature, rainfall and atmosphere. Topographic mapping has also been highly recognized as a comprehensive method to visualize geographical information, such as the reflectance of slope and terrain. Moreover, these concepts have not been applied in the modeling of wireless positioning system. Visualization among wireless infrastructure, large obstacles and spatial distribution of RSS has been studied in this chapter using the multi-layer fuzzy color and topographical map.

6.2 Fuzzy Color Map

This section describes how the multi-layer fuzzy model can be applied. In order to build the fuzzy signal color map, we first need to retrieve the RSS readings in the entire positioning area. The RSS data can retrieved by either manually taking a thorough site-surveying or installing wireless sensor networks to measure the RSS readings.

After we can get all the RSS readings, we need to use the propagation-loss algorithm to approximately interpolate RSS readings among collected locations. The range of the signal strength reading, r are normalized from 0 to 1.

$$\frac{x_{max} - x_{min}}{R_{max} - R_{min}} = \frac{x_{max} - x_i}{R_{max} - R_i}$$

$$x_i = \frac{R_i - R_{min}}{R_{max} - R_{min}} \tag{6.1}$$

where R_{max} is the maximum overall signal strength, R_{min} is the minimum overall signal strength, R_i is the original signal strength, x_{max} is the maximum normalized signal strength and x_{min} is the minimum normalized signal strength, x_i is the value to be normalized.

6.2.1 Fuzzy Membership Function

Using fuzzy logic, the proposed model offers an enhanced LF hyperbolic solution that maps the RSS from a 0 to 1 fuzzy membership function. This approach does not use a numeric value. Instead, it uses fuzzy logic to broadly categorize the RSS as 'strong,' 'normal,' or 'weak.' The normal distribution is used to represent the fuzzy membership functions.

$$P(x) = \frac{1}{\sigma\sqrt{2\pi}} e^{\frac{-(x-\mu)^2}{2\sigma^2}} \tag{6.2}$$

where $P(x)$ is the probability function, x is the normalized RSS, σ is the standard deviation of normalized signal normalized strength in a region, μ is the mean of signal strength in a region. The WLAN network covers the entire campus.

The membership function of term set, $\mu(RSSDensity)$ belongs to the set of [Red,Green,Blue]. Red means the signal strength density is strong; green means the signal strength is normal; and blue means the signal strength density is weak. The fuzzy set interval of blue is [0, 0.5], [0, 1] is green and [0.5, 1] is red.

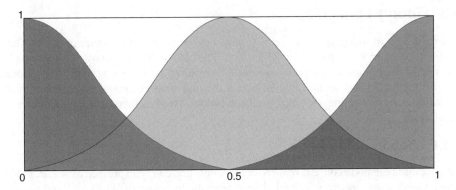

Figure 6.1   The RSS fuzzy membership graph. (A full color version of this figure appears in the color plate section.)

For the blue region,

$$\mu_{blue}\,(0 < x < 0.5) = e^{-150x^2} \tag{6.3}$$

For the green region,

$$\mu_{Green}\,(0 < x < 1) = e^{-150(x-0.5)^2} \tag{6.4}$$

For the red region,

$$\mu_{red}\,(0.5 < x < 1) = e^{-150(x-1)^2} \tag{6.5}$$

Figure 6.1 shows the fuzzy membership function. X-axis represents the normalized signal strength from 0 to 1. The width of membership function depends on the standard deviation of the RSS. The overlap area will be indicated by mixed colors.

6.2.2   Fuzzy Spatio-temporal Cluster

We can use different colored regions to represent the WLAN RSS distribution. Conceptually a spatio-temporal region is defined as follows: Assume that B is a finite set of RSS vectors belonging to a particular color region, where $B = \{b_1...b_n | b_i \in \Re^n\}$, i.e., $b_i \in S, \forall S \in R$, and $\forall S \in [l, u]$, where l is the lower bound of fuzzy interval and u is a upper bound of fuzzy interval. To analyze the distribution surfaces S, there always exists a spatio-temporal mapping, $q : B \rightarrow S$.

$$q(x) = \int_S h(x)b(S)dS \tag{6.6}$$

$$h(x) = \begin{cases} 1 \ x \in S \\ 0 \ x \notin S \end{cases} \tag{6.7}$$

where $h(x)$ is the characteristic function of S, i.e., $b(S)$ is a weight function that specifies a prior on the distribution of surfaces S. We can explicitly define $b(S)$ by propagation-loss algorithm. By (6.2-6.7), the RSS distribution can be illustrated.

6.3 Topographic Map

The basic idea of topographic model is to plot a curve connecting minimum points where the function has a same particular RSS value. The sets of APs are known as topographic line nodes. Topographic line nodes are the APs residing on the topographic lines around contour region. In this section, we introduce the major operations of topographic model including topographic line node measurement, Nelder-Mead method and topographic model generation.

6.3.1 Topographic Node

Each topographic node consists of three components and can be expressed as $< l, d, g >$, in which l represents topographic level, d represents the locations of Wi-Fi received signal, g represents the gradient direction of the RSS distribution. The spatial data value distribution mapped into the (x, y, l) space, where the coordinate (x, y) represents the location and $l = f(x, y)$ describes a function mapping from (x, y) coordinates to level l. The gradient vector g denotes the direction of RSS where to degrade in the space. The gradient vector can be calculated by:

$$g = -f'(x, y) = \left(\frac{\Delta f}{\Delta x}, \frac{\Delta f}{\Delta y} \right)^{T} \tag{6.8}$$

$$S = \frac{B - W}{2} \tag{6.9}$$

6.3.2 Nelder-Mead Method

The Nelder-Mead (NM) method is a commonly used nonlinear optimization algorithm for finding a local minimum of a function of several variables has been devised by Nelder and Mead (Mathews and Fink, 1998). It is a numerical method for minimizing an objective function in a many-dimensional space. We make use of Nelder-Mead method to generate the topographic nodes.

First, we collect the location fingerprint, r with an unknown location (x, y). We define $f(n) = |n - r|$, where n is any location fingerprint. Second, we select three location fingerprints (LFs) to be three vertices of a triangle.

We initialize a triangle BGW and function f is to be minimized. Vertices B, G, and W, where $f(B)$ is the smallest value (best vertex), $f(G)$ is the medium value (good vertex), and $f(W)$ is a largest value (worst vertex). There are four cases when using the NM method. They are reflection, expansion, contraction and shrink. We recursively use NM method until finding the point which is the local minimum (nearest) in B, G, W that are the same value.

The midpoint of the good side is

$$M = \frac{B + G}{2} \tag{6.10}$$

Reflection Using the Point R

The function decreases as we move along the side of the triangle from W to B, and it decreases as we move along the side from W to G. Hence it is feasible that function f takes

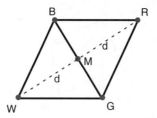

Figure 6.2   Reflection using the point R.

on smaller values at points that lie away from W on the opposite side of the line between B and G. We choose the point R that is obtained by 'reflecting' the triangle through the side BG. To determine R, we first find the midpoint M of the side BG. Then draw the line segment from W to M and call its length d. This last segment is extended a distance d through M to locate the point R (see Figure 6.2). The vector formula for R is

$$R = M + (M - W) = 2M - W \qquad (6.11)$$

Expansion Using the Point E

If the function value at R is smaller than the function value at W, then we have moved in the correct direction toward the minimum. Perhaps the minimum is just a bit farther than the point R. So we extend the line segment through M and R to the point E. This forms an expanded triangle BGE. The point E is found by moving an additional distance d along the line joining M and R (see Figure 6.3). If the function value at E is less than the function value at R, then we have found a better vertex than R. The vector formula for E is

$$E = R + (R - M) = 2R - M \qquad (6.12)$$

Contraction Using the Point C

If the function values at R and W are the same, another point must be tested. Perhaps the function is smaller at M, but we cannot replace W with M because we must have a triangle. Consider the two midpoints C_1 and C_2 of the line segments WM and MR, respectively (see Figure 6.4). The point with the smaller function value is called C, and the

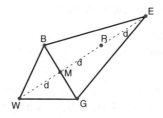

Figure 6.3   Expansion using the point E.

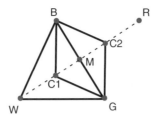

Figure 6.4   Contraction using the point C.

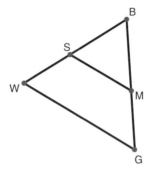

Figure 6.5   Shrink toward B.

new triangle is BGC. Note. The choice between C_1 and C_2 might seem inappropriate for the two-dimensional case, but it is important in higher dimensions.

$$C_1 = \frac{M - W}{2} \tag{6.13}$$

$$C_2 = R - \frac{M - W}{2} \tag{6.14}$$

Shrink toward B

If the function value at C is not less than the value at W, the points G and W must be shrunk toward B (see Figure 6.5). The point G is replaced with M, and W is replaced with S, which is the midpoint of the line segment joining B with W.

6.3.3   Topographic Model Generation

We generate topographic model based on the NM algorithm. We apply NM method to a many-dimensional RSS distribution space problem to simplify the fuzzy color map down to a contour (line-based) map.

First, we select three LFs to be three vertices of a triangle: $\vec{B}$, $\vec{G}$ and $\vec{W}$, where $\vec{B}$ is a location with high RSS (best vertex), $\vec{G}$ is a location with medium RSS (good vertex), and $\vec{W}$ is a location with the low RSS (worst vertex). The location vector of RSS at x_k, y_k use in function, $N(x, y)$. We use the propagation loss algorithm to define $N(x, y)$. There are four

Table 6.1   Nelder-Mead method procedure.

IF f(R)<f(G), THEN Perform Case (i) {either reflect or extend}
ELSE Perform Case (ii) {either contract or shrink}

BEGIN {Case(i)}	BEGIN {Case(ii).}
IF f(B)<f(R) THEN	IF f(R)<f(W) THEN
replace W with R	replace W with R
ELSE	ENDIF
compute E and f(E)	
IF f(E)<f(B) THEN	compute C = (W + M)/2
replace W with E	or C = (M + R)/2 and f (C)
ELSE	IF f(C)<f(W) THEN
replace W with R	replace W with C
ENDIF	ELSE
ENDIF	compute S and f(S)
END {Case (i)}	replace G with M
	ENDIF
	END {Case (ii)}

cases when using the NM method. They are reflection, expansion, contraction and shrink. We recursively use the NM method until finding the point which is the local minimum in $\vec{B}$, $\vec{G}$ and $\vec{W}$ that they are the same value. Table 6.1 summarizes the procedure.

A contour function is then used to plot a curve connecting the minimum points where the function has a same particular value. We normalize the minimum value between 0 and 1, and the contour line is 0.1 in each level.

6.4   Signal Visualization Experiments and Results

This section describes the Wi-Fi signal visualization experiment and results of a Wi-Fi signal in a university campus. The RSS site survey measurements are taken in The Hong Kong Polytechnic University (PolyU) campus. The approximate total area of the campus is 9.34 hectares.

6.4.1   Experimental Setup

Figures 6.6 and 6.7 show a satellite photo of the PolyU campus from Google Earth and a site plan for the same area showing the 27 buildings of interest. In this arrangement, the campus is regarded as having 26 major buildings from Core A to Core Z as well as seven other large buildings, each with WLAN access. The measurements were taken at a total of 27 locations on-campus. A standard laptop computer equipped with an Intel WLAN card and client manager software was used to measure samples of RSS from access points (APs) of PolyU campus. The WLAN card is a chipset inside the laptop.

Each core building is covered by at least 13 APs. The radio frequency (RF) channels of IEEE 802.11b are in the 2.4 GHz band. The received signal sensitivity of the WLAN card limits the range of the RSS to between -93 dBm and -15 dBm. Nevertheless, the highest typical value of the RSS is approximately -30 dBm at one meter from any AP. The sampling

Figure 6.6   Satellite photo of PolyU campus from Google Earth. (A full color version of this figure appears in the color plate section.)

schedule seeks to collect the RSS data every 5 seconds. The sampling will be taken in two periods: between 7:30 am and 9:30 am (not busy) and 4:30 pm and 6:30 pm (busy). The vector of the RSS data at each location forms the location fingerprint with around 20 RSS elements in the vector.

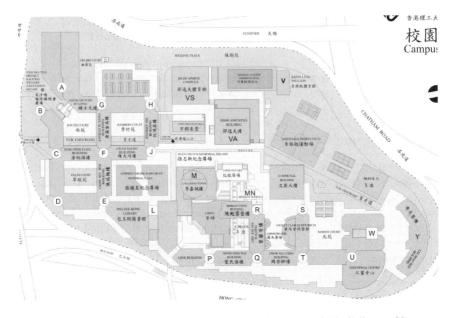

Figure 6.7   Site plan of the PolyU campus identifying the 27 buildings of interest.

6.4.2 Visualization Results

This section describes the Wi-Fi signal visualization results. There are three RSS features to be analyzed, line-of-sight (LOS), the presence of humans, and RSS variation.

Fuzzy Color Layers

Figure 6.8 shows the different RSS color layers. This provides a multi-layer color view of the Wi-Fi signal intensity in different buildings. We can easily identify that there are two major centers of high intensity in the fuzzy color layers.

Fuzzy Color Map and Topographic Map

Figure 6.9 shows that from M core to R core there is low signal strength propagation. On the topographic map in Figure 6.10, the slope of the contour line from M core to R core is steep are the edge, which means that the RSS weakens quickly in the middle from M core to R core due to NLOS effects. The RSS transmission path between buildings can be LOS, partial LOS or shadow where NLOS propagation is possible. For LOS conditions, RSS should fit into lognormal distribution. A multi- story building in a campus area will experience lower signal strengths within tall buildings due to the absence of LOS propagation.

Figure 6.9 shows the effect of LOS in two major clusters of RSS. The two major centers of high intensity are located at F core and S core. The distance from F core to S core is around 600 m. The APs are evenly distributed in the campus, so the signal should be distributed evenly. Between M core (Lee Ka Shing Tower) and R core (Shirley Chan Tower), the RSS distribution is relatively low. The heights of two buildings in M core and R core are around 80 m and 70 m respectively. The distance from M core to R core is around 200 m.

Figures 6.9 and 6.10 show the effect of LOS in two major clusters of RSS. The two major centers of high intensity are located at F core and S core.

A user's presence can affect the mean of the RSS value and the accuracy of location estimation in the system. To investigate this, the same data and collection methods are used as in the previous section. In this case, however, of especial interest are any differences between the measurements taken in the morning and in the afternoon.

Figures 6.11(a) and 6.11(b) show the different RSS patterns when the data was collected in the two different time slots. Figure 6.11(b) shows that the intensity of the red is lower during the busy period. Figures 6.12(a) and 6.12(b) show that in the less busy period the intensity of the red increases from 0.5 to 0.77. In Figure 6.12(a) the topographic region in 0.9 levels is larger than in Figure 6.12(b). The slope in Figure 6.12(b) is smaller than Figure 6.12(a). We can conclude with some confidence that the presence of human bodies will affect the mean of the RSS value. In other words, the RSS will be weaker in busy periods, reducing the accuracy of location estimation.

The accuracy of the tracking system is highly dependent on RSS variation. If the standard deviation of the RSS increases, the accuracy of the tracking system falls. In this it is assumed that to maintain high accuracy the suggested standard deviation of RSS should be under 4 dBm. Other authors in other work have assumed a standard deviation of 2.13 dBm (Taheri *et al.*, 2004). But this work is in particular allowing for the exigencies of real environments, including dense human traffic, and so assumes a large standard deviation.

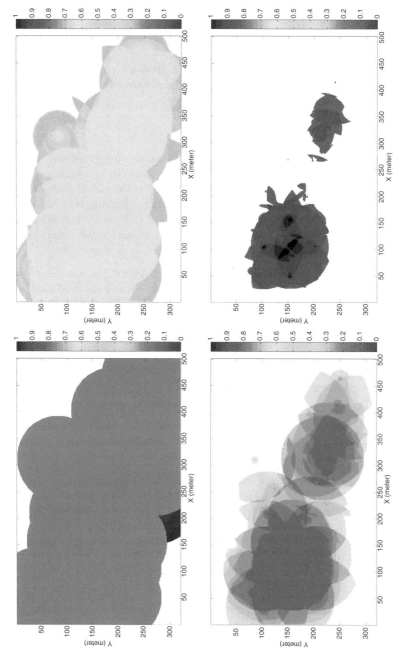

Figure 6.8   Different RSS color layers. (A full color version of this figure appears in the color plate section.)

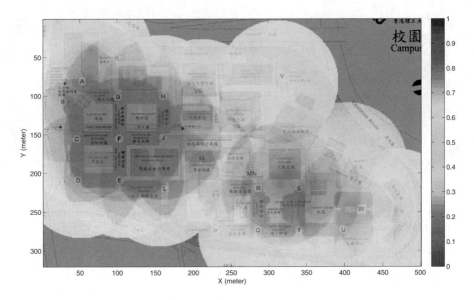

Figure 6.9 Fuzzy RSS distribution with the campus floor plan. (A full color version of this figure appears in the color plate section.)

Accuracy of the tracking system is highly dependent on the RSS path loss exponent α. The path loss exponent α can be varied between 1 and 6 according to the RSS absorbing medium. Path loss exponent α represents the attenuation rate of the RSS. When the path loss exponent α increases, the accuracy of the tracking system becomes higher. Therefore, high path loss exponent is an easy way to track the target.

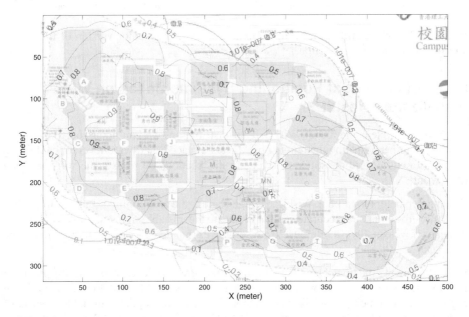

Figure 6.10 Topographic RSS distribution with the campus floor plan. (A full color version of this figure appears in the color plate section.)

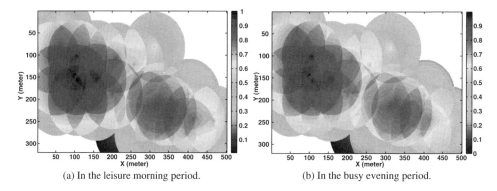

(a) In the leisure morning period.   (b) In the busy evening period.

Figure 6.11   RSS distribution in fuzzy analytical model. (A full color version of this figure appears in the color plate section.)

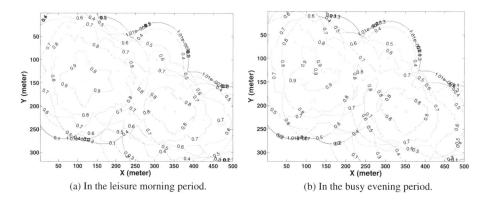

(a) In the leisure morning period.   (b) In the busy evening period.

Figure 6.12   RSS distribution in topographic model. (A full color version of this figure appears in the color plate section.)

6.5   Refinement of Positioning Systems Based on Wi-Fi Visualization Result

After we get a clear picture of Wi-Fi signal distribution, we can adjust our positioning systems by either adjusting the number of APs to be placed (hardware-level) or calibrating the positioning algorithm (software-level). In this section, we look at the software-level refinement.

When a person enters a poor Wi-Fi coverage region, the positioning accuracy drops dramatically. To solve the inaccurate estimation due to the unstable signal, intuitively, it would be better if we can always reference from some APs that are stable enough to estimate the position.

Figure 6.13 shows the flowchart which illustrates the process of each step of the the proposed positioning approach. As shown in the flowchart, we first test the stability of signal by the Signal-to-Interference-plus-Noise-Ratio (SINR). Then if the signal is unstable, instead of picking K-nearest fingerprints, we only pick the fingerprints from the AP in the red region. By using this method, we can slightly improve the distance error by 7% in average.

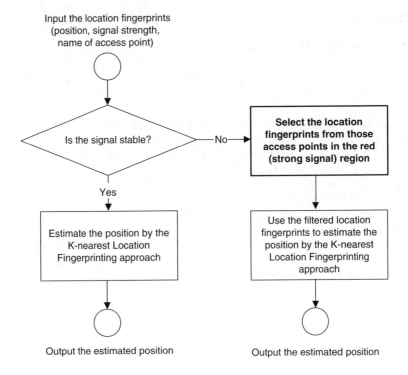

Figure 6.13   Flowchart of the proposed positioning approach.

Chapter Summary

In this chapter, we have introduced two modeling methods: fuzzy color map and topographic map. Fuzzy color map provides detailed, multi-color layers to visualize the Wi-Fi received signal strength (RSS) where red means the signal strength density is high, green means the signal strength is medium and blue means the signal strength density is low. A topographic map serves as a quick reference and line-based visualization tools where the RSS is denser and clustering different RSS in different topographic levels. Finally the visualization result helps to provide necessary information to refine the positioning system in both hardware and software level.

References

Bahl, P., Padmanabhan, V., and Balachandran, A. (2000). A software system for locating mobile users: design, evaluation, and lessons. *Microsoft Online Research Document*.

Budianu, C., Ben-David, S., and Tong, L. (2006). Estimation of the number of operating sensors in large-scale sensor networks with mobile access. *IEEE Transactions on Signal Processing*, pp. 1–13.

Chan, C. L., Baciu, G., and Mak, S. (2009). Fuzzy topographic modeling in WLAN tracking analysis. *International Conference on Fuzzy Computation, ICFC*, pp. 17–24.

Jan, R. and Lee, Y. (2003). An indoor geolocation system for wireless LANs. *International Conference on Parallel Processing Workshops*, pp. 29–34.

Kaemarungsi, K. and Krishnamurthy, P. (2004a). Modeling of indoor positioning systems based on location fingerprinting. *The 23th IEEE International Conference on Computer Communications, INFOCOM*, 2, pp. 1012–1022.

Kaemarungsi, K. and Krishnamurthy, P. (2004b). Properties of indoor received signal strength for WLAN location fingerprinting. *The 1st International Conference on Mobile and Ubiquitous Systems: Networking and Services, MOBIQUITOUS*, pp. 14–23.

Kwon, J., Dundar, B., and Varaiya, P. (2004). Hybrid algorithm for indoor positioning using wireless LAN. *The IEEE 60th Vehicular Technology Conference, VTC*, 7, pp. 4625–4629.

Mathews, J. and Fink, K. (1998). *Numerical Methods Using MATLAB*. New York: Simon & Schuster.

Swangmuang, N. and Krishnamurthy, P. (2008). Location Fingerprint Analyses toward Efficient Indoor Positioning. *The 6th IEEE International Conference on Pervasive Computing and Communications*, pp. 101–109.

Taheri, A., Singh, A., and Emmanuel, A. (2004). Location fingerprinting on infrastructure 802.11 wireless local area networks (WLANs) using Locus. *The 29th IEEE International Conference on Local Computer Networks*, pp. 676–683.

Wong, W., Ng, J., and Yeung, W. (2005). WLAN positioning with mobile devices in a library environment. *The 25th IEEE International Conference on Distributed Computing Systems Workshops*, pp. 633–636.

Part II

Outdoor Positioning Systems

Chapter 7

Introduction of Global Positioning System

*In most people's vocabularies, design means ve-
neer. It's interior decorating. It's the fabric of the
curtains of the sofa. But to me, nothing could be
further from the meaning of design. Design is the
fundamental soul of a human-made creation that
ends up expressing itself in successive outer layers
of the product or service.*

Steve Jobs
1955–2011

Chapter Contents

► History of GPS

► Functions of GPS

► GPS components

► Types of GPS receiver

► Sources of errors in GPS

► Coordinate system in GPS

Global Positioning System (GPS) is a satellite-based radio navigation system developed and operated by the US Department of Defense (DOD). GPS is free for anyone since 1996. GPS utilizes the constellation of at least 24 Earth orbit satellites that transmit the signals, and then the GPS receivers collect those signals from satellites to calculate the user position anywhere in the Earth. In this chapter, we look at the history, functions, components and the coordinate system of GPS.

Introduction to Wireless Localization: With iPhone SDK Examples, First Edition. Eddie C.L. Chan and
George Baciu.
© 2012 John Wiley & Sons Singapore Pte. Ltd. Published 2012 by John Wiley & Sons Singapore Pte. Ltd.

7.1 History of GPS

The design idea of GPS was based on ground-based radio navigation systems in the early 1940s. Originally, those system were used in World War II. The GPS made a breakthrough when the first man-made satellite which was called 'Sputnik' was created in 1957. At that time, a team of US scientists researched the radio transmission of Sputnik and found that the signal could be transmitted far away from Earth. They thought that the satellite could be utilized for navigation and positioning on the global Earth. The first test of positioning and navigator on the Earth was carried out at the United States Department of Defense in 1960. US Navy and Air Force used a constellation of satellites to achieve an accurate positioning. This was the first satellite based navigation system which was called 'Transit.'

In 1973, the US Department of Defense used US Navy and Air Force systems to create a GPS for the military and civilian uses. This system is called 'NAVigation System with Timing And Ranging' (NAVSTAR). NAVSTAR provides the signal from which can be calculated the accurate position, time, and velocity on the Earth. There was a program called 'NAVSTAR JPO' which was managed by the US military. The program helped to manage, maintain, and control the GPS components.

1996 was a milestone year in GPS development. The US President Bill Clinton issued the directive which declared that GPS could be used by either military users or civilian users. And the Interagency GPS Executive Board (IGEB) was established and responded to manage the system as a US national asset.

Selective Availability (SA)

Selective Availability is developed by US government for adding error into GPS signal (inside the Coarse/Acquisition (C/A) code). Although civilian users have been allowed to use GPS since 1996, the US government added errors to affect the accuracy of positioning. This accuracy reduction system is called 'Selective Availability' (SA). The reason for development of SA is that the US government wanted to prevent the enemy from getting an extremely precise position of important buildings and weapons guidance. SA error is generated by cryptographic algorithms by which the random numbers are added to the satellites' clock to make them inaccurate. The normal users did not have permission to decrypt the SA, so the distance error could be up to 100 meters.

Anti-Spoofing (AS)

Anti-Spoofing (AS) allows decryption of precision GPS coordinates. This is implemented by interchanging the Precision code (P code) with a classified Y code (from the US military GPS receivers) in GPS Signal. (This will be further covered in Chapter 8.)

On May 2, 2000, US President Clinton ordered SA to be turned off, and the accuracy of civilian users improved after that day. SA has been permanently deactivated due to the broad distribution and worldwide use of the GPS system. US military has other ways to prevent terrorist attacks using the GPS data for locating important buildings with homemade remote control weapons.

7.2   Functions of GPS

Nowadays, GPS has become a very popular and easy-to-use system and is very low-cost for recreational users. The development of GPS has had a great impact on the world. There are four basic and major functions of GPS: positioning in coordinates, distance measurement, velocity measurement, and accurate time measurement.

GPS Positioning Features

Positioning is the major function of GPS. GPS receivers receive satellite signals and read the GPS string data. In general, the GPS string consists of the position given as latitude/longitude, the height above sea level, number of available satellites and a checksum. There are at least of 24 GPS orbiting satellites above the Earth. Users can receive at least 6 satellite signals from anywhere. The satellites continuously use FM waves to broadcast information from space back to the Earth. The GPS signal packages from satellites includes the time sent from satellites, the orbital information of satellites, and the status of satellites. When the GPS receivers collect the signals from satellites, then the received GPS signal packages are used to calculate the position of the receivers.

Distance Measurement

Although Earth is ellipsoid, GPS data can be used to measure the distance between two points, or between the receiver and the destination. For example, GPS can measure distance between New York and San Francisco. We will talk about it further in Section 7.7.

Velocity Measurement

With the tracking ability of GPS, it can also help to calculate the speed of movement of an object. It estimates the time of arrival according to the velocity of an object. This measurement is very useful in estimating the time and to control speed in marine navigation.

Accurate Time Measurement

GPS can also act as a universal timepiece, since the satellites orbit around the Earth. The trajectory of a satellite is generally set to circulate around the Earth every 24 hours. The atomic clocks in satellites are precise to within a billionth of a second. This features of GPS can be applied to synchronize the time among receivers precisely.

7.3   Components of GPS

GPS is a US space-based radio navigation system that consists of three major segments of usage. They are space segment, control segment, and user segment. Figure 7.1 shows the relationships between these three segments in GPS.

7.3.1   Space Segment

Space segment is the portion of a satellite system which consists of the satellites, their power and navigation systems, and communication equipment. There are at least 24 GPS

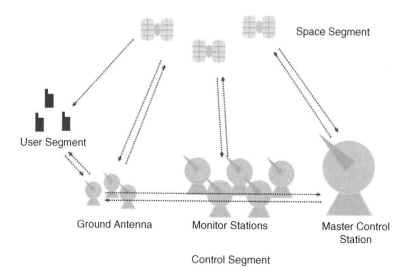

Figure 7.1   Three segments in GPS.

satellites about 12 000 miles above the Earth. In the Spring of 2008, the number of GPS satellite constellations increased to 32 satellites which has highly improved the accuracy of positioning. The satellite's life is approximately 10 years. The weight of the satellite is about 2000 pounds, and it is about 17 feet across with its solar panels extended. The satellites run around the Earth in six orbital planes, and each satellite is oriented at an angle of approximately 55 degrees to the equator and approximately 60 degree right ascension of the ascending node.

7.3.2   Control Segment

The control segment is a portion of a communication system that coordinates the operation of its key components, including master control station, monitor stations and ground antenna.

Master Control Station

Master Control Station (MCS) is for controlling and uplinking data to the satellites. The data includes clock-correction, atmospheric, almanac data and ephemeris data of each unique satellite.

The Consolidated Satellite Operations Center is a MCS which is located at the US Air Force Space Command Center at Schriever Air Force Base in Colorado Springs, Colorado.

Monitor Station

Monitor Stations (MS) is for monitoring the satellites. The function of MS is to collect the satellite signal data and then transmit it to the MCS for evaluation. MS acts as a powerful GPS receiver which can track the satellites and collect the satellite signal precisely. There are five MSs on the Earth. They are located at Schriever Air Force Base, Hawaii, Kwajalein Atoll, and Diego Garcia, and the Ascension islands.

Ground Antenna

Ground antenna helps to uplink the data to satellites, which is controlled by the MCS. Also, it can collect the telemetry data from the satellites. There are three ground antennas located at Ascension, Diego Garcia, Kwajalein Atoll.

7.3.3 User Segment

User segment is the portion of a communication system that interfaces the systems from users, including the receivers, processors, and antennas which receive the satellite signals, and then calculate the precise position, velocity and time. User segment can be applied in lots of areas such as military, surveying, transportation, and recreation.

7.3.4 Ground Segment

The ground segment is typically for Differential GPS (DGPS) to improve the performance of reference stations and provides more precise ephemerides for the localization. It is used in DGPS which measures the satellite pseudoranges and real pseudoranges to calibrate the pseudoranges in the user segment.

7.4 Types of GPS Receivers

There are many types of GPS receivers on the market with a wide range of price. The price of a GPS receiver generally depends on performance. There are four factors that can affect the performance of GPS receivers as listed below:

- ▶ The channel design of the receiver – the number of channels are designed to affect the performance of finding satellites. Single channel receiver means the receiver only can find one satellite at a time. Nowadays, receivers supporting 12 channels are very common in the market.

- ▶ Size of storage – if the size of storage in the receiver can be larger, the receiver can collect and store bigger number of coordinates, user-defined attributes, and the raw data from the satellite.

- ▶ Supporting multiple coordinate systems – generally, a common GPS receiver only displays the geodetic coordinates. But an expensive receiver might support other types of coordinate system, so that it can display the position using other coordinate systems.

- ▶ Battery life – GPS receiver has longer battery life that can obtain more data. Note that the extreme temperature may affect the performance of the GPS receiver battery.

Having discussed the factors affecting the performance of GPS receivers, let's look at different types of GPS receivers used in different areas.

Recreational Receiver

Recreational GPS receivers are designed to acquire a location fix quickly without error correction. This is the cheapest GPS receiver and easy buy anywhere at a cost of around US$100. In order to meet the lower cost requirements of outdoor enthusiasts, most recreational units give a position ignoring the quality of GPS signals they receive. For recreational purposes, this is perfectly adequate and suitable. For example, hikers can generally find where they are within 10 meters, which is far enough.

Mapping Grade Receivers

Mapping grade receivers are designed to collect a large number of raw data and coordinates. It is more precise than the recreational receiver. The raw data can be used by Geographic Information System (GIS) to produce a map. The price of a mapping grade receiver is expensive.

Survey Grade Receiver

The survey grade receiver is a very powerful type of GPS receiver which is used for surveying the landscape. Precision is up to a few centimeters and it has a large amount of storage space for data collection.

US Military/Government Receiver

The US government receiver has one special feature which can receive the precision code (P-code) from the satellites. It calculates more accurate positioning and navigation to a precision of a few centimeters from the P-code.

7.5   Sources of Errors in GPS

Generally, there are six types of errors that can adversely affect positioning accuracy and precision. They are ephemeris errors, satellite clock errors, receiver errors, atmospheric errors, multi-path interference and user equivalent range error.

7.5.1   Ephemeris Errors

Ephemeris errors are generated by the GPS signal which includes the incorrect location of satellite. Generally, the MSC updates the current location of satellites and predicts the location of satellites in the next period. Then, it uplinks to the satellites. However, the ephemeris errors occurred while prediction of the satellites' location not matching with the actual location. The GPS receiver uses the almanac data to calculate the traveling distance of the satellite. If errors exist in almanac data, then the calculation will also contain some errors which will lead to inaccurate positioning. The ephemeris errors can be up to 8 meters. To increase the accuracy and precision, the post processing of almanac data should be at later periods. However, it is not suitable for the real-time positioning tracking.

7.5.2 Satellite Clock Errors

Satellite clock errors refer to satellite clock bias. When there are biases in the clock measurement, it may cause a delay of around 1 to 2 nanoseconds. Since the satellite clock measurement is used to calculate the distance to the satellite, if the clock is inaccurate, then it affects the sub-sequence calculation. There might be around 3 m distance error on average when a bias satellite clock is used. The distance error caused by the clock bias can derived as:

$$D = t \cdot c \tag{7.1}$$

where D is the distance error caused by clock bias, t is the time difference and c is the light of speed (299 792 458 m/s). For example, when there is a 1 nanosecond of clock bias, the distance error is 3 meters ($0.000000001 * 299\,792\,458 \approx 3$).

We can calibrate the satellite clock error using the synchronization process driven by the ground control stations. The ground control stations can determine precisely the solution of second-order polynomials which is a set of equations to calculate the time drift. The time drift message is then broadcast to the satellites within 1 millisecond. This time synchronization among the satellites' clocks takes within 20 nanoseconds. Also, the time synchronization between the satellite standard time and the Coordinated Universal Time (UTC) takes within 100 nanoseconds. UTC is the primary time standard by which the world regulates clocks and time. During this short period of time, the satellite clocks are in random time drift which causes unpredictable satellite clock errors.

7.5.3 Receiver Errors

Receiver errors refer to the clock bias of the GPS receiver. The atomic clock in the satellite is very precise which is up to 11 decimal places. In contrast to some receivers, they are less precise and only measure the time up to 6 decimal places. In this case, when the receiver gets the GPS signals, the time is rounded up. This roundup error leads to inaccurate positioning. But the impact of this is small which only causes less than 1 meter error. Moreover, there might have been some noises during reception, such as GPS signal noise, multi-path delays which increase the receiver clock errors.

7.5.4 Atmospheric Errors

Atmospheric errors are made when the signals pass though the ionosphere and troposphere to the GPS receiver. The ionosphere and troposphere are 150 km and 10 km above the Earth surface.

Ionospheric Delays

The ionosphere is the first layer that the GPS signals pass though the atmosphere. The ionosphere contains free electrons which can delay the signals to penetrate which leads to an error in calculation. When the electromagnetic (signal) wave generated by the satellite penetrates to the ionospheric layer, it will refract by the free electrons.

There are three major factors that can affect the degree of signal refraction. They are sunspot activity, different time period of a day, and the Earth geometry.

▶ Sunspot activity – the higher levels of sunspot activity cause larger delay of signals that leads to larger errors.

▶ Different time period of a day – there is a huge range of difference of error between the daytime and nighttime. In daytime, the ionosphere could delay the signals and cause an error of about 40–60 meters; however, at nighttime, the error distance is only 6–12 meters.

▶ Earth geometry – the ionosphere is unstable in the polar and equatorial zones which leads to larger delays and errors. However, the ionosphere is much more stable in temperate zones, and then the delay is smaller. As a result, the error is smaller and is just about 2 to 5 meters.

There is a solution to improve the ionospheric error. Users could use the dual-frequency receiver which can receive the signals in L1 band and L2 band at the same time. (L1 and L2 bands of GPS signal are introduced in Chapter 8.) When the receiver collects the signals in L1 and L2 bands continuously, it calibrates the difference between both signals and increases the precision up to 5 meters.

Tropospheric Delays

After passing through the ionosphere, the EM wave then passes though the troposphere. In this layer, the temperature, humidity and air pressure can affect the formation of air which may lead to a delay in the signals. The dry and wet air components are the causes of tropospheric delays which could refract the signals in the troposphere. The distance error caused by the dry component is:

$$D_{component} = (2.27 \cdot 0.001) \cdot P_{surface} \qquad (7.2)$$

where $D_{component}$ is the distance error caused by the dry component, and $P_{surface}$ is the pressure on surface which is in millibar (mb). For instance, when the pressure of a region in troposphere is 2000 mb, the distance error is $(2.27 \times 0.001) \times 2000 = 4.54$ meters.

For the wet component, it is difficult to estimate error since many factors such as temperature, humidity, altitude, and angle of the signal path become unstable and unclear. Up till now, scientists still have no general equations to calculate the wet component factor. Anyway, the tropospheric delay has little impact on the signal which usually causes less than 1 meter distance error.

7.5.5   Multipath Interference

The radio waves from the satellites cannot penetrate solid objects such as mountain and buildings. However, the radio waves are deflected by these solid objects. These deflections create constructive and destructive interferences of the radio waves that lead to multipath interference. If there are a lot of objects which have large reflective surfaces, then the distance error can be up to 15 meters or more. To resolve this error, we can collect the signal

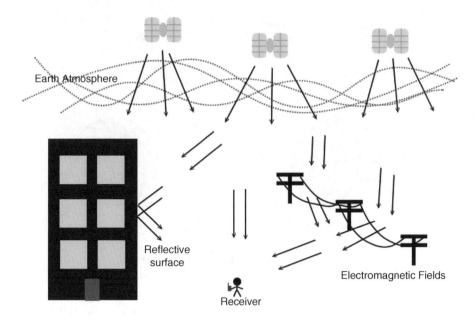

Figure 7.2   Sources of errors.

values in a period of time and average them, so that the multipath effect can be minimized. New model of GPS receivers generally includes the reduction of multipath effect.

Figure 7.2 show three possible sources of errors, multipath interference, ionospheric and tropospheric delays.

7.6   Precision of the GPS

There are two main components that determine the precision of the GPS: Geometric Dilution of Precision (GDOP), User Equivalent Range Error (UERE).

7.6.1   Geometric Dilution of Precision (GDOP)

Geometric Dilution of Precision (GDOP) or Dilution of Precision (DOP) is the geometric effect of the spatial relationship of the satellites relative to the user. In surveying terms, it is the 'strength of figure' of the trilateration position computation. GDOP varies rapidly with the time since the satellites are moving. As such, the precision in the pseudorange of the satellite translates to a corresponding component in each of the four dimensions of position measured by the receiver (i.e. x, y, z, and t). GDOP can be defined as:

$$GDOP = \sqrt{\sigma_{east}^2 + \sigma_{north}^2 + \sigma_{up}^2 + \sigma_{overall}^2 + (c \cdot \sigma_{time})^2} \cdot \left(\frac{1}{\sigma_{overall}} \right) \qquad (7.3)$$

where σ_{east}, σ_{north} and σ_{up} are standard deviations of positions in east, north, up in the Earth coordinate system and σ_t is the standard deviation of time measurement.

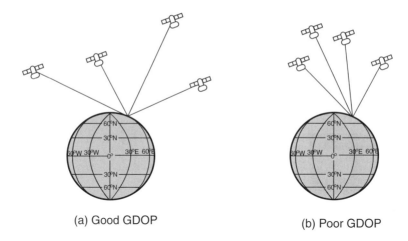

(a) Good GDOP   (b) Poor GDOP

Figure 7.3   Low and high GDOP values.

Using Equation (7.3), we can know when visible GPS satellites are close together in the sky, the GDOP value is high. When they are far apart, the GDOP value is low. A low GDOP value represents a better GPS positional precision due to the wider angular separation between the satellites used to calculate a GPS unit's position. Figure 7.3 shows the cases of low and high GDOP values.

GDOP can be expressed as a number of separate measurements: positional (3D), horizontal, vertical and time.

Position Dilution of Precision (PDOP)

Position Dilution of Precision (PDOP) refers to the horizontal and vertical spatial relationships of the satellites relative to the user. PDOP combines with two dilutions of precisions. They are the Horizontal Dilution of Precision (HDOP) and the Vertical Dilution of Precision (VDOP). The ideal positions of satellites are evenly distributed in the sky. And the poor distribution is that satellites clustered together.

Horizontal Dilution of Precision (HDOP)

$$HDOP = \sqrt{\sigma_{east}^2 + \sigma_{north}^2} \cdot \left(\frac{1}{\sigma_{overall}}\right) \tag{7.4}$$

Vertical Dilution of Precision (VDOP)

$$VDOP = \sigma_{up} \cdot \left(\frac{1}{\sigma_{overall}}\right) \tag{7.5}$$

$$PDOP = \sqrt{\sigma_{east}^2 + \sigma_{north}^2 + \sigma_{up}^2} \cdot \left(\frac{1}{\sigma_{overall}}\right) \tag{7.6}$$

7.6.2 User Equivalent Range Error (UERE)

UERE is the error of the individual range measurement to each satellite. UERE also varies between different satellites, atmospheric conditions, and receivers. The absolute range precision is largely dependent on which code (C/A or P-Code) is used to determine positions. If UERE coupled with the geometrical relationships of the satellites during the position determination, this results give an uncertainty of 3D ellipsoid in all three coordinates. To address this error, error propagation techniques in statistics field are applied to define the confidence level of the estimated position.

7.7 Coordinate Systems on the Earth

In this section, we introduce the coordinate system used in GPS. The coordinate system should pinpoint the specific location in the n-dimensional space. In our Earth, we use coordinate (x, y, z) to represent the location in a three-dimensional space. However, the coordinate system used in the Earth is more complex.

Geodetic Coordinate System

In fact, the Earth is not a perfect sphere and is approximately an ellipsoid. It is slightly flattened at the poles and wider at the equator. A geodetic coordinate system is a three-dimensional coordinate system used in GPS. It uses latitude, longitude and height to represent the specific location on the Earth surface.

Latitude

Latitude is the angle from the surface to the equatorial plane, which is the center of the Earth. The 0 degree of latitude is on the equator, and 90 degrees north is at the North Pole and 90 degrees south is at the South Pole.

Longitude

Longitude is the angle east or west of a reference meridian between the two geographical poles to another meridian that passes through an arbitrary point. The 0 degree of longitude is at the Royal Observatory, Greenwich in England. This reference line is also called the Prime Meridian. The location to the east from the Prime Meridian is 180 degrees east and the location to the west from the Prime Meridian is 180 degrees west.

Height

The height represents the position above or below the ellipsoid (or sea level in more precise definition).

As shown in Figure 7.4, latitude (Φ) and longitude (λ) are angular coordinates expressed in degrees. Positive latitudes are in north of the equator (usually in the northern hemisphere), while negative latitudes are in south. Positive longitudes are in east of Prime Meridian while negative longitudes are in west.

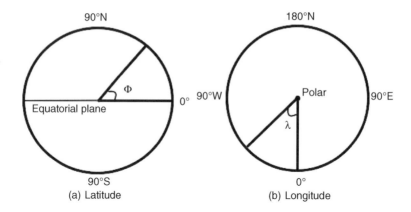

Figure 7.4   Geodetic coordinates.

Each degree of latitudes and longitude is subdivided into 60 arc minutes; each arc minute is divided into 60 arc seconds. In the iPhone and most programming API, latitudes and longitudes are expressed as a decimal fraction.

$$\text{DecimalMinutes} = \text{Degrees, Minutes} + (\text{Seconds}/60) \tag{7.7}$$

$$\text{DecimalDegrees} = \text{Degrees} + (\text{Minutes}/60) + (\text{Seconds}/3600) \tag{7.8}$$

For example, the latitude and longitude of Hong Kong are equivalent in the following expressions:

Degrees, minutes, seconds

Latitude: $22°20'45\text{N}$

Longitude: $114°11'30\text{E}$

Decimal minutes

Latitude: $22°20.75'\text{E}$

Longitude: $114°11.5'\text{E}$

Decimal degrees

Latitude: 22.3458 N

Longitude: 114.1917 E

Lines of Latitude and Longitude

An equal distance lies between each line of latitude; thus lines of latitude are often referred to as parallels. Figure 7.5 shows the earth's grid which consists of lines of latitude and longitude. For simplicity, we assume the Earth to be a sphere. Each length of arc-degree of

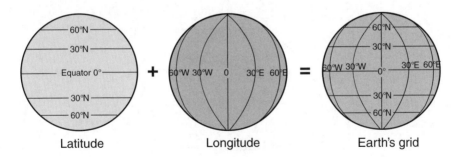

Figure 7.5   Lines of latitude and longitude.

north–south latitude difference could be calculated by:

$$\frac{\pi}{180} \times \text{ radius of the Earth} \tag{7.9}$$

The radius of the Earth is approximately equal to 6369 km or 3957 miles. An equal distance lies between each line of latitude is approximately equal to 69 miles or 111 km. An equal degree of rotation lies between lines of longitude, which are sometimes referred to as meridians. However, since the circumferences of parallels vary with latitude, the distance between lines of longitude varies from 69 miles at the equator to zero degrees at the poles. Each length of an arc-degree of east–west longitude difference could be calculated by:

$$22°22'45 \times \text{ radius of the Earth} \times cos\,(\text{latitude}) \tag{7.10}$$

Great Circle Distance

The great circle distance or orthodromic distance is the shortest distance between any two points on the surface of a sphere measured along a path on the surface of the sphere (as opposed to going through the sphere's interior). Because spherical geometry is rather different from ordinary Euclidean geometry, the equations for distance take on a different form. The distance between two points in Euclidean space is the length of a straight line from one point to the other. On the sphere, however, there are no straight lines. For example, the distance from A and B in an Earth is actually not a straight line but is an arc length as shown in Figure 7.6. In non-Euclidean geometry, straight lines are replaced with Geodesics. Geodesics on the sphere are the great circles (circles on the sphere whose centers are coincident with the center of the sphere).

Let $A\,(\phi_1, \lambda_1)$ and $B\,(\phi_2, \lambda_2)$ be the latitude and longitude of point A and point B respectively. The great circle distance between two points could be calculated as follow:

$$d\,(A, B) = 2r \arcsin \sqrt{\sin^2\left(\frac{\phi_1 - \phi_2}{2}\right) + \cos\phi_1 \cos\phi_2 \sin^2\left(\frac{\lambda_1 - \lambda_2}{2}\right)} \tag{7.11}$$

where r is the radius of the Earth. For example, we can make use of the above equation to calculate the distance between Hong Kong and London. The latitude and longitude of

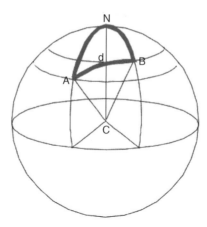

Figure 7.6   Great circle distance.

London is 52N, 0. During calculation, you may need to change degree to radian by $\pi/180$. The distance is 9602 km.

Planar Coordinate System

As the geodetic coordinates system gives a representation of coordinates in the 3D model, it is more complex to measure distance between two points on the map. For example, it is inconvenient and difficult to measure the distance between two objects on the 3D sphere. Therefore, a planar coordinate system (Cartesian coordinate system) is introduced to represent the location of the Earth in a 2D coordinate system. The transformation process of mapping information from three to two dimensions is called 'Map Projection.'

Chapter Summary

This chapter has covered the fundamental GPS components, types of GPS receiver, GPS errors and the GPS coordinate system. In the next chapter, we will cover more about GPS signal and its positioning algorithms.

Chapter 8

Study of GPS Signal and Algorithms

We didn't build the Mac for anybody else. We built it for ourselves.

Steve Jobs
Feb. 1985

Chapter Contents

- ▶ What are the GPS signals?

- ▶ What is navigation message format?

- ▶ Introducing the modernized GPS signals

- ▶ Introducing the position algorithms used in GPS

- ▶ Calculating the user velocity

S atellites broadcast GPS radio signals to the Earth's surface. GPS signal includes ranging and navigation messages which are used to measure the distance from the Earth to satellites. With GPS signals, we can apply algorithms to determine the location and synchronize time of the user. In this chapter, we first look at the GPS signal and later introduce the algorithms used in GPS.

8.1 GPS Signals

The basic design of GPS signal contains three types of codes: Coarse Acquisition code (C/A-code), Precision code (P-code) and the navigation message. There are two carrier signal bands, L1 and L2. L1 band is at a frequency of 1575.42 MHz with 19 cm wavelength band and L2 band is at a frequency of 1227.60 MHz with 24.4 cm wavelength. The data rate in both L1 and L2 band is in 50 bits per second. GPS satellites transmit the codes in these two carrier signal bands.

Introduction to Wireless Localization: With iPhone SDK Examples, First Edition. Eddie C.L. Chan and George Baciu.
© 2012 John Wiley & Sons Singapore Pte. Ltd. Published 2012 by John Wiley & Sons Singapore Pte. Ltd.

8.1.1 Coarse Acquisition Code

C/A code is found on the L1 band. The C/A code is a 1023 bit deterministic sequence called pseudorandom noise code (also pseudorandom binary sequence) (PN or PRN code). And the code sequence repeats every 1 millisecond. The carrier transition rate is at 1.023 Mbps with the physical distance between binary transitions is 293 meters.

The power spectrum of C/A code is 2.046 MHz of the null-to-null bandwidth. C/A code is a pseudorandom code which is generated by a known algorithm. The code contains the time according to the satellite clock when the signal was transmitted. Because each satellite has their unique C/A code, the GPS receiver can identify which satellites the signals came from. C/A code is freely available to the public.

8.1.2 Precision Code

P code is found on both L1 and L2 band. And the code sequence repeats every 267 days. Compared with C/A code, P code is better for more precise positioning. The carrier transition rate is at 10.23 Mbps with the physical distance between binary transitions being 29.3 meters. However, P code was encrypted by a process known as 'Selective Availability (SA)', which was introduced by the US Department of Defense. The purpose of encryption of P-code was to add errors to affect the accuracy of positioning to civilian users. Until May 2, 2000, SA is permanently deactivated and P code is now not encrypted.

Figure 8.1 shows the GPS signals in L1 and L2 bands.

8.1.3 Navigation Message

Navigation message is transmitted in the L1 band, and the transmission rate is 50 bps. It includes the information of the satellites. There are three components that are included in the navigation message. First, the message includes the general satellite information such as satellite time, status of satellite. Second is orbital information for satellite calculation, and third is the almanac information which contains ephemeris data for all the satellites.

Almanac is a set of data that every GPS satellite transmits, and it includes information about the state (health) of the entire GPS satellite constellation, and coarse data on every satellite's orbit. Ephemeris data is very precise orbital and clock correction for each satellite which is necessary for precise positioning. The GPS Control Segment generates a new almanac and

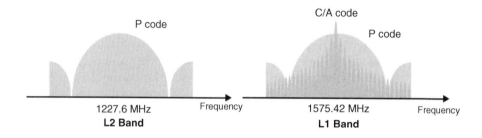

Figure 8.1   The GPS signals in L1 and L2 bands.

ephemeris data every day and sends them to each satellite with the next scheduled navigation data upload. In the following section, we introduce the navigation message format briefly.

8.1.4 Navigation Message Format

The navigation message format is divided into five subframes, and each subframe contains 300 bits. Because the transmission rate is 50 bps, so that it needs to take 6 seconds to transmit one subframe. In each subframe, there are ten 30-bit words. Each word in subframes has reserved the last six bits for parity checking and the parity checking is used for checking the bit error during demodulation. Hamming code technique is applied to parity checking. Hamming code is a linear error-correcting code which could detect up to two simultaneous bit errors, and correct single-bit errors.

The transmission of the navigation message starts with the order from subframe 1 to subframe 5. As subframe 4 and 5 are larger, containing 25 pages, they are required to transmit several times in order to complete the broadcasting. For example, if it is the first time to transmit the five subframes, page 1 of subframe 4 and 5 are transmitted first. Subsequently, page 2 of subframe 4 and 5 are transmitted and so on. It will take a total of 25 times for transmitting and broadcasting all pages of both subframe 4 and 5.

There is a common characteristic of each subframe. In each subframe, the first two words are telemetry data and a handover word. Telemetry data (TLM) is the first word in each subframe which identifies the beginning of the subframe. The first part of TLM is assigned with the fixed value of 1001011 which is in 8-bit pattern. The second part of the TLM contains another 14 bits of data which is for the authorized users.

The second word in the subframe is the handover word (HOW). It contains several parameters in the word which allows the device to handover the C/A code tracking and P code tracking. HOW contains the subframe number to identify which current subframe is broadcasted at each transmission. Since each subframe takes 6 seconds to transmit, HOW defines the GPS time-of-week modulo six seconds to each corresponding subframe. There are 2 flag bits for alert indicator and anti-spoofing indicator. For alert indicator flag, it presents the signal accuracy. When the flag is set, it means that the signal accuracy is poor. For the anti-spoofing indicator, it shows whether the anti-spoofing is activated or not.

Figure 8.2 shows the general idea of the navigation message format of each subframe. In the following paragraphs, let's briefly introduce each subframe.

Subframe 1

Subframe 1 consists of three parts: GPS transmission week number, clock corrections, and the satellite accuracy and health information.

▶ GPS transmission week number – the cycle of each satellite is 1024 weeks. 1024 is the standard number of weeks of each satellite. If the number of weeks started a cycle on January 5, 1980, the first cycle would be finished on August 22, 1999. The second cycle will be finished on April 2019 and so on. The transmission week number is captured by the GPS receiver to keep track of the satellites' periods.

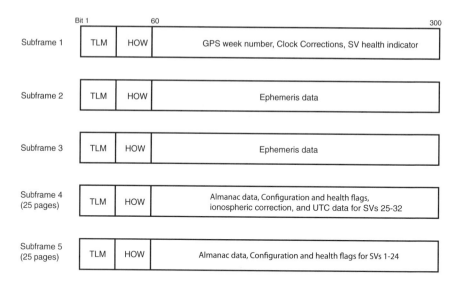

Figure 8.2   Navigation message format.

▶ Clock corrections – there are 10 bits numbers which refer to the issue of data and clock (IODC). IODC is used for identifying the current set of navigation data. IODC is unique and the receiver can check the IODC to find out the change of the current navigator dataset. The IODC will be repeated in a seven-day cycle. Thus, subframe 1 provides the satellite clock correction terms for precise ranging which helps for the synchronization between the time of satellite clock and the time of transmission signal from the satellite. Group delay (GD) corrections are applied for the reduction of the ionospheric errors. GD is used by the single frequency receiver. Since the clock correction terms refer to the timing of P code in the L1 and L2 band, GD helps the receiver to measure the ionospheric errors by using the linear combination of the dual frequency of P code in L1 and L2 bands.

▶ Satellite accuracy and health information – this consists of four components: user range accuracy indicator, satellite health indicator, L2 code indicator, and L2 P-data flag. The user range accuracy indicator can estimate 1 sigma range errors caused by space and control segments. The satellite health indicator is a 6-bit indicator to check the signal contains error and indicate whether the satellite works properly. The L2 code indicator identifies whether the C/A code or P code is active on the L2 band or not. And the L2 P-data flag identifies how the navigation data is modulated in P-code on L2 band.

Subframe 2

Subframe 2 includes the ephemeris data which is used in the satellite location measurement. The first part of subframe 2 are orbital elements known as Keplerian elements which are used for locating the satellite. Keplerian elements are the inputs to a standard mathematical model of spacecraft orbits. Second, there is a fit interval flag and an age of data offset (AODO) term. The fit interval flag checks whether the orbital elements based on a nominal

4-hour curve or a longer interval. And the AODO refers to a navigation message correction table which has been included in the GPS navigation data since 1995. The final part of subframe 2 is an issue of data ephemeris which contains the 8 least significant bits used for monitoring the changes of the orbit elements.

Subframe 3

Subframe 3 is similar to subframe 2. It also contains the ephemeris data for the orbital elements. And it also includes the issue of data ephemeris for checking the changes of orbital elements.

Subframe 4

Subframe 4 contains data for the satellites number 25–32 which includes the almanac data, the configuration and health flags, ionospheric corrections, and the UTC time for the satellite. Subframe 4 are broadcasted with subframe 5 together.

An almanac data is in from page 2 to page 5 and from page 7 to page 10 in subframe 4. This is a coarse data for assisting the receiver to determine the position of the satellites, so that the GPS receiver can obtain the data easily.

Second, the page 13 in subframe 4 is the Navigation Message Correction Table (NMCT) which is applied in the range correction. Third, the page 18 in subframe 4 includes ionospheric corrections which help for the GPS receiver to synchronize the UTC time to GPS time. GPS time is retrieved by the single frequency receiver. Fourth, the page 25 in subframe 4 is used for the configuration and health flags of the satellite. Finally, the remaining pages are reserved for the future use.

Subframe 5

Subframe 5 is similar to subframe 4 which also includes the almanac data, and health data flags. Since there are 32 satellites orbit on the Earth, subframe 5 contains the almanac data for the satellites 1–24. First, page 1 to page 24 in subframe 5 is the almanac data, and the almanac data assists the receiver to find out the satellites, position. Second, the health data flags is in page 25. Each of subframe 4 and subframe 5 contains 25 pages.

8.2 Modernized GPS Signals

In the previous section, we've mentioned the traditional GPS signals. There are four new GPS signals which have been added to satellites since 2006. They are L1 civil (L1C) signal, L2 civil (L2C) signal, L5 signal, and M code signal. L1C, L2C and L5 are the new signals for the civil usage, and M code is the new signal for military usage. L1C is still under development. In the following paragraphs, we briefly describe these four new signals used in the modern GPS system. Figure 8.3 shows the modernized GPS signals in L1, L2 and L5 Bands. Table 8.1 summarizes the bandwidth and frequency of modernized GPS signals.

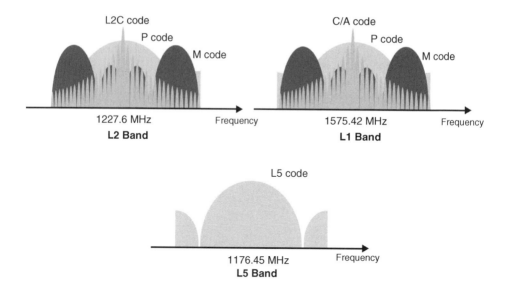

Figure 8.3   The modern GPS signals in L1, L2 and L5 bands.

8.2.1   L2 Civil Signal (L2C)

L2C signal is at the 1227.6 MHz. Its power spectrum is similar to the C/A code which is 2.046 MHz null-to-null bandwidth. Although they are in the similar power spectrum, they have different codes. L2C signal is fixed up with the two different PN codes.

The first PN code refers to civil moderate code. It repeats its sequence in every 10 230 chips. The second code is civil long code. It repeats in every 767 250 chips. In other words, it repeats 75 times for each cycle. Both of the codes are generated at 511.5-k chips per second.

8.2.2   L5 Signal

L5 signal is at the 1176.45 MHz which is similar to L2C encoding. It repeats the period in every 1 millisecond. L5 signal is encoded by I5 and Q5 PN codes. Neuman-Hofman (NH) synchronization codes are used in I5 and Q5 codes. For I5 PN code, there is a 10-symbol

Table 8.1   Bandwidth and frequency of modernized GPS signals.

GPS signals types	Bandwidth (MHz)	Frequency (MHz)
C/A code	1.023	1575.42
P code in L1	10.23	1575.42
P code in L2	10.23	1227.6
L2C in L2	10.23	1227.6
M code in L1	15.345	1575.42
M code in L2	15.345	1227.6
L5	10.23	1176.45

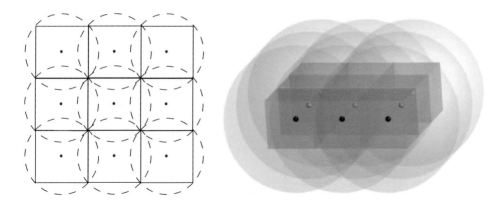

Figure 2.9   Square tessellation.

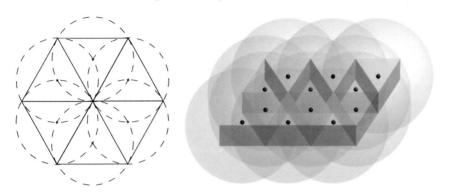

Figure 2.13   Triangular tessellation.

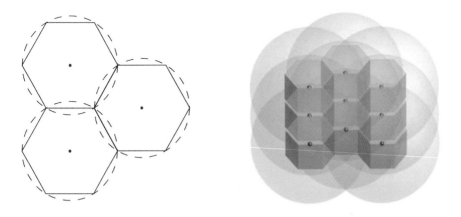

Figure 2.14   Hexagonal tessellation.

Introduction to Wireless Localization: With iPhone SDK Examples, First Edition. Eddie C.L. Chan and George Baciu.
© 2012 John Wiley & Sons Singapore Pte. Ltd. Published 2012 by John Wiley & Sons Singapore Pte. Ltd.

Figure 4.9   Customized Wi-Fi positioning system in iPhone.

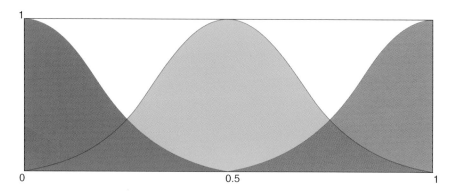

Figure 6.1   The RSS fuzzy membership graph.

Figure 6.6   Satellite photo of PolyU campus from Google Earth.

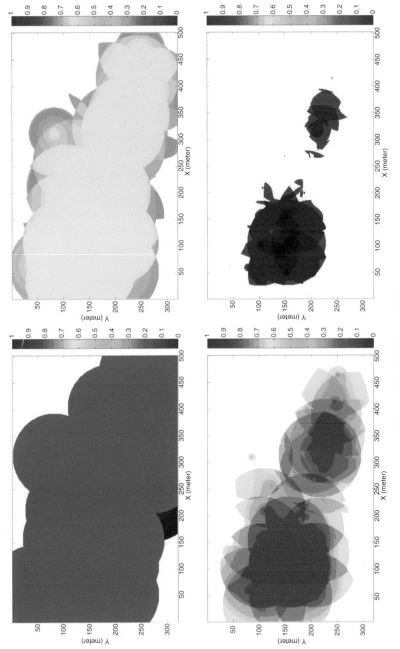

Figure 6.8   Different RSS color layers.

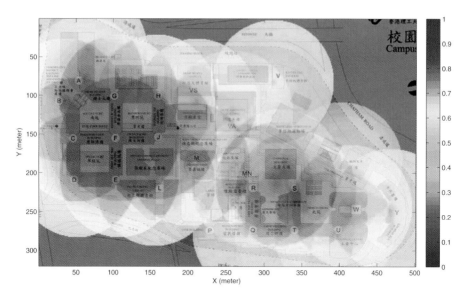

Figure 6.9   Fuzzy RSS distribution with the campus floor plan.

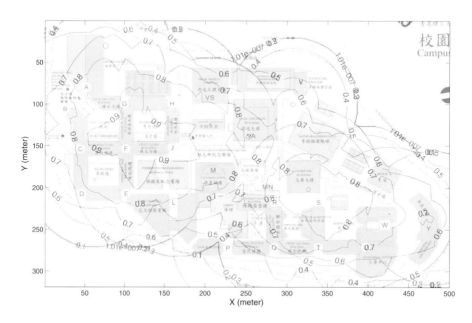

Figure 6.10   Topographic RSS distribution with the campus floor plan.

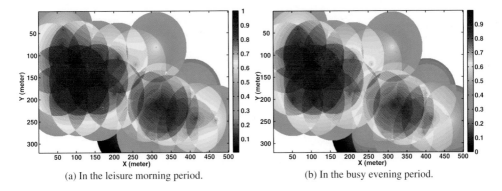

(a) In the leisure morning period.

(b) In the busy evening period.

Figure 6.11   RSS distribution in fuzzy analytical model.

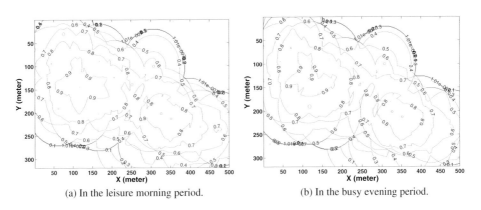

(a) In the leisure morning period.

(b) In the busy evening period.

Figure 6.12   RSS distribution in topographic model.

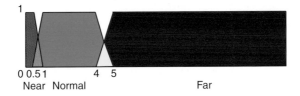

Figure 11.6   Fuzzy membership graph for distance.

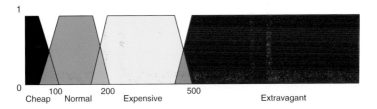

Figure 11.7   Fuzzy membership graph for price.

Figure 11.8   Layout of Tourist Guide.   Figure 11.9 Selection in searching function.

Figure 11.12   Using Google Maps to show the guiding trajectory to target restaurant.

Figure 12.15   Geo-tagged friend objects on map view.

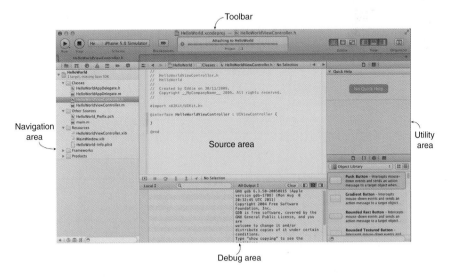

Figure A.8   Xcode environment for developing iPhone applications.

(a) Standard view.

(b) Hybrid view.

(c) Localize to US with hybrid view.

(d) Zoom to Hong Kong with hybrid view.

(e) Zoom to Hong Kong with standard view.

(f) Press Alt key to zoom in US.

Figure B.26   Different types of view.

(a) Zoom to San Francisco.

(b) Zoom to a Hong Kong City.

(c) Zoom to China.

Figure B.29   Zoom to different area.

NH code which is generated every 10 milliseconds. For Q5 PN code, there is a 20-symbol NH code.

8.2.3   M Code

M code is the new GPS signal for the military usage. M code can be found in L1 and L2 bands. The design of M code is to replace P code in time. M code has a much powerful security than P code and supports anti-jam resistance. In addition, M code has better performance in positioning tracking and data demodulation and it is compatible with the C/A code and P code.

M code is a binary offset carrier (BOC) signal which is generated by two parameters. BOC is a square sub-carrier modulation. The first parameter is in square-wave sub-carriers, and it has 10×1.023 MHz. The second parameter refers to the generator code chipping rate. There are 5 generators in M code and each generator operates at 1.023 MHz. Thus, the bandwidth of the M code is $10 \times 1.023 + 5 \times 1.023 = 15.345$ MHz.

8.2.4   L1 Civil Signal (L1C)

L1 civil signal (L1C) is based on the L1 carrier frequency which seeks to maximize interoperability with the GPS signal. L1C is designed to have center frequency of 1575.42 MHz which is the pre-eminent GPS frequency for a variety of reasons, including the extensive use of GPS C/A code, the lower ionospheric error at L1 band relative to lower frequencies, spectrum protection of the L1 band, and the use of this same center frequency by GPS.

L1C design leverages including advances in signal design knowledge, improvements in receiver processing techniques, developments in circuit technologies, and enhancements in supporting services such as communications. The L1C design has been optimized to provide superior performance, while providing compatibility and interoperability with other signals in the L1 band.

We've learnt so far about the components of GPS signals. In the next section, we will learn about positioning algorithms used in GPS.

8.3   GPS Absolute Point Determination

This section covers the GPS absolute point determination. Absolute point positioning involves the use of only one single passive receiver at the user's location to collect data from multiple satellites. The received ranging and navigation messages in GPS signals are applied with the trilateration algorithm to determine the user's position.

8.3.1   Trilateration Algorithm

Trilateration positioning algorithm uses trigonometry geometry to compute the location of an object. In a 2D environment, the algorithm requires three satellites' GPS signal.

Let's use a daily life example to illustrate the concept about the trilateration positioning. Assume a person (let's say, Eddie) would like to identify his location and seek his way in the map. He noticed that he was 5 kilometers away from the Victoria Peak in Hong Kong.

Figure 8.4   Calculating the possible location using one single reference point.

Figure 8.4 shows a possible location (the area of circle) that Eddie may be located according to this piece of information.

Later, he realized that he was also 8 kilometers away from Aberdeen. Another circle was drawn in Figure 8.5 in which two circles intersects at two points.

He saw the street plate and he noticed that he was 4 kilometers away from the Central. A third circle was drawn in Figure 8.6 in which three circles intersect at a point where Eddie should be there.

In this example, we deal with a 2D space situation. The trilateration technique can also be applied in a 3D space as well, but we are dealing with spheres instead of circles. As we have three unknowns (x, y, z) in a 3D coordinate, we need to have four sphere equations to determine the exact user's position. Specifically, a GPS receiver should find four (or more) satellites to measure the distances among GPS receiver and satellites, then we can make use of these measurements to calculate the user location in the Earth. If the receiver can only find three visible satellites, then it can use an imaginary sphere to represent the Earth in calculations but there will be no altitude information.

As solving a set of equations is not an easy task, it would be better if we can further simplify the above problem. Assume we only deal with 2D space positioning and three visible

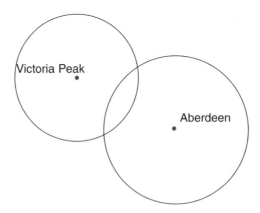

Figure 8.5   Calculating the possible location using two reference points.

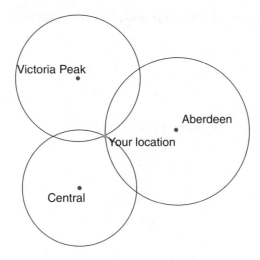

Figure 8.6   Calculating the possible location using three reference points.

satellites are identified, if we put the first satellite as the origin of the coordinate system, so the coordinate of first satellite is $(0, 0)$. Then, we put the second satellite on the y-axis, and the coordinate of second satellite is $(c, 0)$. Finally the third satellite is on the xy-plane which is related to the first and second satellite, the coordinate of third satellite is (a, b). The object location is (x, y). The radius of the three satellites is $r1$, $r2$, and $r3$. Figure 8.7 illustrates the new coordinate system based on the above descriptions.

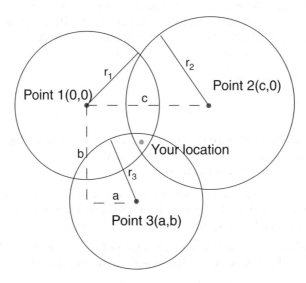

Figure 8.7   Shifting the coordinate system.

As a result, we are able to calculate the possible location as follows:

$$r_1^2 = x^2 + y^2 + z^2 g^\circ$$
$$r_2^2 = (x - c)^2 + y^2 + z^2 \tag{8.1}$$
$$r_3^2 = (x - a)^2 + (y - b)^2 + z^2$$

From Equation (8.1) , we can get the following expression:

$$y = \frac{r_2^2 - r_1^2 - c^2}{-2c}$$
$$x = \frac{r_3^2 - r_1^2 - b^2 - a^2 + 2by}{-2a} \tag{8.2}$$
$$z = \pm\sqrt{r_1^2 - x^2 - y^2}$$

If the z is the square root of a negative number, this represents there is no intersection points. If three satellites intersects at a point, and then the z value should be equal to zero. If there is one positive and one negative value in z, these two z values represent either one of these two points in the object location.

Now that we have this basic understanding of trilateration, we will talk further about what pseudorange is.

8.3.2   What is Pseudorange?

Pseudorange is the distance between a satellite and a receiver. It is calculated by the time difference between the starting transmission time of the satellite (according to an atomic clock in satellite) and the receiving time of the GPS receiver (according to the receiver's clock). A pseudorange observation is equal to the true range from the satellite to the user plus delays due to satellite/receiver's clock biases and other effects. The satellite's atomic clock is highly precise, whereas the GPS receiver's clock is less precise. Figure 8.8 shows the pseudorange observation from satellite to receiver.

Pseudorange (pr) is calculated as:

$$pr = (t - t_s) c \tag{8.3}$$

where t is the time when the signal is received from the satellite according to the receiver's clock. t_s is the time when the signal transmits from the satellite according to the atomic clock in the satellite. c is the speed of light.

For example, there is a signal which takes 10 milliseconds to transmit from the satellite to the receiver. We can calculate the distance between them. The speed of light in a vacuum is 299 792 458 m/s, therefore, the pseudorange is 299 792 meters.

There are two components about the measured time t from the receiver. They are the true reception time tr and the clock bias b. The time measurement from the receiver is:

$$t = tr + b \tag{8.4}$$

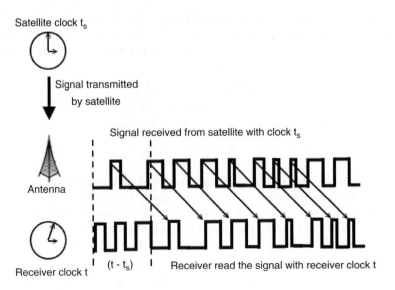

Figure 8.8   Pseudorange observation from a satellite to a receiver.

The satellite's clock time has true transmission time tr_s and its clock bias b_s. The time measurement from the satellite is:

$$t_s = tr_s + b_s \tag{8.5}$$

The next step is to substitute the true time and clock bias into the pseudorange equation. And we have the following equation:

$$pr(tr) = ((tr + b) - (tr_s + b_s)) c$$
$$pr(tr) = (tr - tr_s) c + cb - cb_s \tag{8.6}$$
$$pr(tr) = pr_{true}(tr, tr_s) + cb - cb_s$$

where $pr_{true}(tr, tr_s)$ is the pseudorange from the satellite to receiver without any clock bias. c represents the speed of light in the atmosphere, and it ignores the refraction.

8.3.3   Determining the Location

Using Pythagoras' theorem , we can have the below equation for solving $pr_{true}(tr, tr_s)$.

$$pr_{true}(tr, tr_s) = \sqrt{(x_s(tr_s) - x(tr))^2 + (y_s(tr_s) - y(tr))^2 + (z_s(tr_s) - z(tr))^2} \tag{8.7}$$

where (x_s, y_s, z_s) is the position of the satellite. The GPS receivers calculate the satellite position from the GPS signals which know the satellite position (x_s, y_s, z_s) and the clock bias is b_s .

Generally, the transmission time takes approximately 0.007 seconds when the satellite's signal starts to transmit, until the receiver collects the signal. The satellite range can change about 60 meters in this short transmission period. If we use the reception time to calculate, the distance error could be tens of meters. Therefore, we need to include the clock bias of the satellite in the calculations. 'Light time equation' is introduced to adjust the time error. It is an iterative equation which uses the reception time t. 'Light time equation' is defined as:

$$tr_s(0) = tr = (t - b)$$
$$tr_s(1) = tr = -\frac{pr_{true}(tr, tr_s(0))}{c}$$
$$tr_s(2) = tr = -\frac{pr_{true}(tr, tr_s(2))}{c}$$

$$....$$

(8.8)

The receiver calculates the satellite position continuously using the Keplerian-type element until the computed range converges. Equation (8.8) starts the iteration from the true reception time, t and the receiver's clock bias, b. However, it is not easy to measure the receiver's clock bias precisely. Fortunately, such bias is only within a few milliseconds that only leads to a few meters' error distance. Therefore, we can neglect the receiver's clock bias.

We use 3D coordinate (x,y,z) to represent a point in the 3D Earth space. In this case, we need to find four unknowns during each calculations. Unknowns are receiver position (x,y,z) and the receiver's clock bias b. Using the above Equation 8.7, we can define the pseudoranges to each satellite as:

$$pr_{true} + (cb_s - cb) = \sqrt{(x - x_s)^2 + (y - y_s)^2 + (z - z_s)^2}$$

(8.9)

where (x, y, z) is the unknown point where we need to determine, (x_s, y_s, z_s) is the position of the satellite, and the clock bias from the satellites is b_s, and pr_{true} is the pseudorange from the satellite to receiver without any clock bias. To solve four unknowns (x, y, z, b), at least 4 satellites should be used in solving the equations. Derived from Equation (8.9), we have a set of equations:

$$pr_1 = \sqrt{(x - x_1)^2 + (y - y_1)^2 + (z - z_1)^2} + cb - cb_1$$
$$pr_2 = \sqrt{(x - x_2)^2 + (y - y_2)^2 + (z - z_2)^2} + cb - cb_2$$
$$pr_3 = \sqrt{(x - x_3)^2 + (y - y_3)^2 + (z - z_3)^2} + cb - cb_3$$
$$pr_4 = \sqrt{(x - x_4)^2 + (y - y_4)^2 + (z - z_4)^2} + cb - cb_4$$

(8.10)

where (x_1, y_1, z_1), (x_2, y_2, z_2), (x_3, y_3, z_3), (x_4, y_4, z_4) are the position of four satellites respectively and b_1, b_2, b_3, b_4 are the clock biases of four satellites respectively. Figure 8.9 shows the example of localization using four satellites in a 3D space.

The unknowns (x, y, z, b) in the above equations can be solved by three methods, closed-form solutions, iterative techniques based on linearization, and Kalman filter.

8.3.4 Determining the Location Using Linearization

In this section, we introduce the linearization method to solve the set of equations (8.10). First, we denote (x, y, z) for the true location, and let $(\hat{x}, \hat{y}, \hat{z})$ be the approximate location. Therefore, the true location is adjusted with the approximate location $(\Delta x, \Delta y, \Delta z)$.

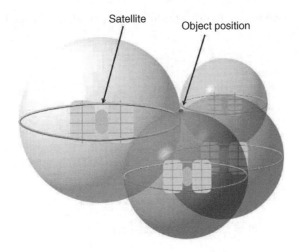

Figure 8.9   Positioning using four satellites in a 3D space.

A function $f(x, y, z, b)$ is derived from Equation (8.9):

$$f(x, y, z, b) = \sqrt{(x - x_s)^2 + (y - y_s)^2 + (z - z_s)^2} + cb - cb_s \qquad (8.11)$$

As mentioned before, the true location could be simplified again by the approximated adjustment. So we now have:

$$
\begin{aligned}
x &= \hat{x} + \Delta x \\
y &= \hat{y} + \Delta y \\
z &= \hat{z} + \Delta z \\
b &= \hat{b} + \Delta b
\end{aligned}
\qquad (8.12)
$$

As a result, we can have:

$$f(x, y, z, b) = f\left(\hat{x} + \Delta x, \hat{y} + \Delta y, \hat{z} + \Delta z, \hat{b} + \Delta b\right) \qquad (8.13)$$

Next, we expand Equation (8.13) by using a Taylor series:

$$
\begin{aligned}
f(x, y, z, b) = {}& f\left(\hat{x} + \Delta x, \hat{y} + \Delta y, \hat{z} + \Delta z, \hat{b} + \Delta b\right) \\
& + \frac{\partial f(\hat{x},\hat{y},\hat{z},\hat{b})}{\partial \hat{x}} \Delta x + \frac{\partial f(\hat{x},\hat{y},\hat{z},\hat{b})}{\partial \hat{y}} \Delta y \\
& + \frac{\partial f(\hat{x},\hat{y},\hat{z},\hat{b})}{\partial \hat{z}} \Delta z + \frac{\partial f(\hat{x},\hat{y},\hat{z},\hat{b})}{\partial \hat{b}} \Delta b + \dots
\end{aligned}
\qquad (8.14)
$$

After eliminating nonlinear terms using the first-order partial derivatives, the partial derivatives are as follows:

$$
\begin{aligned}
\frac{\partial f(\hat{x},\hat{y},\hat{z},\hat{b})}{\partial \hat{x}} &= -\frac{x_s - \hat{x}}{\hat{a}_s} \\
\frac{\partial f(\hat{x},\hat{y},\hat{z},\hat{b})}{\partial \hat{y}} &= -\frac{y_s - \hat{y}}{\hat{a}_s} \\
\frac{\partial f(\hat{x},\hat{y},\hat{z},\hat{b})}{\partial \hat{z}} &= -\frac{z_s - \hat{z}}{\hat{a}_s} \\
\frac{\partial f(\hat{x},\hat{y},\hat{z},\hat{b})}{\partial \hat{b}} &= c
\end{aligned}
\tag{8.15}
$$

where

$$
\hat{a}_s = \sqrt{(\hat{x} - x_s)^2 + (\hat{y} - y_s)^2 + (\hat{z} - z_s)^2}
\tag{8.16}
$$

After that, we substitute Equations (8.15) and (8.16) into Equation (8.17):

$$
pr_{true} = \widehat{pr}_{true} - \frac{x_s - \hat{x}}{\hat{a}_s}\Delta x - \frac{y_s - \hat{y}}{\hat{a}_s}\Delta y - \frac{z_s - \hat{z}}{\hat{a}_s}\Delta z + cb
\tag{8.17}
$$

As a result, we finished the linearization of Equation (8.17) with the adjustment variables Δx, Δy, Δz, and Δb. Then we rearrange Equation (8.18) with the adjustment variables on right yields, like this:

$$
\widehat{pr}_{true} - pr_{true} = \frac{x_s - \hat{x}}{\hat{a}_s}\Delta x + \frac{y_s - \hat{y}}{\hat{a}_s}\Delta y + \frac{z_s - \hat{z}}{\hat{a}_s}\Delta z - cb
\tag{8.18}
$$

Then we simplify the equation with the variables below:

$$
\begin{aligned}
\Delta pr &= \widehat{pr}_{true} - pr_{true} \\
k_{xs} &= \frac{x_s - \hat{x}}{\hat{a}_s} \\
k_{ys} &= \frac{y_s - \hat{y}}{\hat{a}_s} \\
k_{zs} &= \frac{z_s - \hat{z}}{\hat{a}_s}
\end{aligned}
\tag{8.19}
$$

Because k_{xs}, k_{ys}, and k_{zs} are the direction cosines of the unit vector pointing from the approximate location to the sth satellite. We denote the satellite into a matrix, for sth satellite, we have

$$
k_s = \left(k_{xs}, k_{ys}, k_{zs}\right)
\tag{8.20}
$$

After we simplify the variables, Equation (8.20) is rewritten as:

$$
\Delta pr_s = k_{xs}\Delta x + k_{ys}\Delta y + k_{zs}\Delta z - c\Delta b
\tag{8.21}
$$

For solving the four adjustments Δx, Δy, Δz, and Δb, a set of linear equations is formed using four different satellites:

$$
\begin{aligned}
\Delta pr_1 &= k_{x1}\Delta x + k_{y1}\Delta y + k_{z1}\Delta z - c\Delta b \\
\Delta pr_2 &= k_{x2}\Delta x + k_{y2}\Delta y + k_{z2}\Delta z - c\Delta b \\
\Delta pr_3 &= k_{x3}\Delta x + k_{y3}\Delta y + k_{z3}\Delta z - c\Delta b \\
\Delta pr_4 &= k_{x4}\Delta x + k_{y4}\Delta y + k_{z4}\Delta z - c\Delta b
\end{aligned}
\tag{8.22}
$$

And then we transform the set of equations (8.22) into a matrix form, like this:

$$
\Delta p = \begin{bmatrix} \Delta pr_1 \\ \Delta pr_2 \\ \Delta pr_3 \\ \Delta pr_4 \end{bmatrix}, \; T = \begin{bmatrix} k_{x1} \; k_{y1} \; k_{z1} \; 1 \\ k_{x2} \; k_{y2} \; k_{z2} \; 1 \\ k_{x3} \; k_{y3} \; k_{z3} \; 1 \\ k_{x4} \; k_{y4} \; k_{z4} \; 1 \end{bmatrix}, \; \Delta q = \begin{bmatrix} \Delta x \\ \Delta y \\ \Delta z \\ -c\Delta b \end{bmatrix}
\tag{8.23}
$$

As a result, we have:

$$
\Delta p = T\Delta q
\tag{8.24}
$$

We can find the adjustment Δq by:

$$
\Delta q = T^{-1}\Delta p
\tag{8.25}
$$

When the adjustment variables Δx, Δy, Δz, and Δb are calculated, we could find out the true position using Equation (8.23). The accuracy of using the linearization method depends on the adjustment which is within close vicinity of the linearization point. If the adjustment exceeds the expectation value, the approximate location will be updated by a new estimate of pseudorange based on the calculated location (x, y, z). On the other hand, the localization of the true position is affected by errors such as multipath errors, measurement noise, and those errors which are translated into the component in Δq.

The ranges from the clock bias lead the estimated location not to intersect in a single point. Therefore, the clock is required to be synchronized and adjusted until a single point converges. Figure 8.10 shows the effect of the clock bias. In Figure 8.10, when the outer

Figure 8.10 Synchronization of the clock bias.

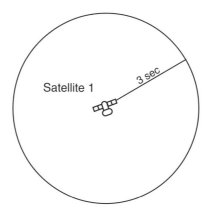

Figure 8.11   Signal traveling time from Satellite 1 to the GPS receiver.

boundaries of the satellites are the calculation of position with clock bias, there is no intersection at a single point. After synchronizing the clock time, we can find the position which is at the single point.

When the satellites transmit the signal to the GPS receiver, the receiver can get the transmission time from the satellites. The time difference between the reception time and the starting transmission time is applied to calculate the distance between the satellites and the user. As a result, we can use trilateration algorithm to determine the receiver's location. Let's briefly explain it with an example. Suppose Satellite 1 transmits the signal to a GPS receive which takes 3 seconds to reach the GPS receiver. Figure 8.11 shows the signal traveling time from Satellite 1 to the GPS receiver.

However, the receiver's clock is a conventional quartz clock, which is relatively less accurate than the atomic clocks in satellites. There is a clock bias in calculating the distance from the satellite to the receiver. If the clock error is 0.01 second, 3000 km error distance will be obtained. Thus, if the precision is ten meters, the clock bias should be around 0.00000003 second. Figure 8.11 shows the signal traveling time. Synchronization of the clock bias is applied to reduce the error distance.

To simplify the explanation, we only assume two satellites have a single clock bias value of 0.5 second in the receiver's clock. So the actual received transmission times from Satellite 1 and 2 are 3.5 seconds and 5.5 seconds respectively. Figure 8.12 shows the actual transmission time of Satellite 1 and 2.

Satellite 3 is added and used in the calculation. The actual received transmission time from Satellite 3 is 4.5 seconds. Figure 8.13 shows the actual transmission time of Satellite 1, 2 and 3 and the actual location of point A.

As can be seen in the figures, ignoring clock bias can increase the distance error. If we ignore the clock bias, we will have distance error to calculate the intersections at point B where the actual location of receiver is at point A. To resolve this error issue, the receiver's clock should be synchronized with the satellites' clock, such that three circles intersect at Point A and we can calculate the actual location.

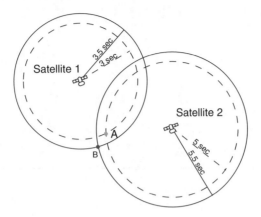

Figure 8.12   Transmission times of Satellite 1 and 2.

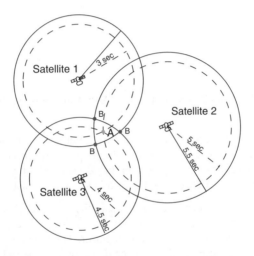

Figure 8.13   Actual location is at Point A.

8.4   Calculating User Velocity

One of GPS functions is to calculate the user velocity. There are several methods to calculate the user velocity. A simple method is that the user velocity is calculated by the approximate derivative of the user position, like this:

$$\dot{p} = \frac{dp}{dt} = \frac{p(t_2) - p(t_1)}{t_2 - t_1} \tag{8.26}$$

where $\dot{p}$ is the user velocity, $p(t_2)$ and $p(t_1)$ are the user positions at the time t_2 and t_1, and $p(t_2) - p(t_1)$ means that the traveling distance of the user from the time t_2 and t_1. The velocity is calculated in a certain time interval. In this calculation, some factors are not included such as the acceleration or deceleration of the movement.

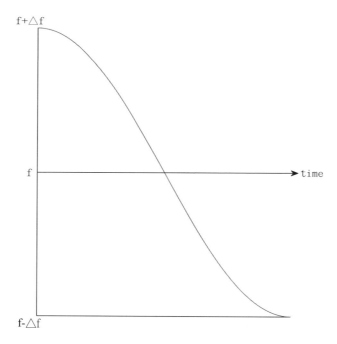

Figure 8.14   Received Doppler frequency by user at rest on the Earth's surface.

Nowadays, modern GPS receivers uses an advanced method using the carrier-phase mea-surements for calculating the user velocity. This requires the precise estimation of the Doppler frequency of the satellite signals. Figure 8.14 shows the curve of the received Doppler frequency by the user at rest on the Earth's surface against the time.

The received frequency value increases, as the satellite approaches nearer to the receiver. The value decreases as it move away from the receiver. There is a Doppler shift which is caused by the motion of the satellite related to the user. The Doppler shift is zero when the satellite is in the closest position to the user. At that moment, the radial component of the velocity of the satellite related to the user is zero. The Doppler equation is defined as:

$$rf = tf\left(1 - \frac{(v_r \cdot l)}{s}\right) \tag{8.27}$$

where rf represents the receiver frequency, tf represents the frequency of transmitted satel-lite signal, c represents the speed of the propagation, v_r represents the velocity difference, l represents the radial component of the relative velocity by the line of sight to the satellite, and then a dot product $(v_r \cdot l)$ denotes the velocity vector respect to the satellite and user.

The velocity difference v_r can be defined as:

$$v_r = v - \dot{p} \tag{8.28}$$

where v is the velocity of the satellite, and $\dot{p}$ is the user velocity, and both of them refer to a common Earth-Centered and Earth-Fixed (ECEF) co-ordinate system. The Doppler offset

Δf from the relative motion can be defined as:

$$\Delta f = rf - tf \tag{8.29}$$

And then substituting Equations (8.27) and (8.28) into (8.29), we have:

$$\Delta f = -tf \frac{(v - \dot{p} \cdot l)}{s} \tag{8.30}$$

For example, at the L1 band, the maximum Doppler frequency for the user on the Earth is approximately 4 kHz, when it has a maximum line of sight velocity in 800 m/s.

There are many methods using the Doppler frequency to calculate the user velocity. We will introduce one of those common methods in the following paragraphs. First, we assume the true location and adjustment (Δx, Δy, Δz) have been estimated (the true location has been estimated and showed in the previous section). Since we calculate the user velocity in a 3D space, the user velocity $\dot{p}$ is denoted as the vector ($\dot{x}$, $\dot{y}$, $\dot{z}$), and b is the clock bias.

As we use at least four or more satellites for positioning, and there may exist ith satellite, we can put Equation (8.28) into (8.27). We have:

$$rf_i = tf_i \left(1 - \frac{(v_i - \dot{p}) \cdot l_i}{s}\right) \tag{8.31}$$

where rf_i is the ith actual satellite frequency collected by the receiver and tf_i transmitted frequency from the ith satellite. As mentioned before, the control stations will send a message to correct the error offsets. As a result, the receiver could get the actual transmitted frequency from the satellite:

$$tf_i = f_0 + \Delta tf_i \tag{8.32}$$

where f_0 is the nominal transmitted satellite frequency such as L1 band and L2 band, and Δtf_i is the corrected frequency of the ith satellite.

We can denote f_i for the estimated signal frequency from ith satellite, and there is a clock bias b for this estimated signal frequency. The unit of b is in second. If the receiver's clock runs faster than the satellite's clock, then b should be a position value. As the estimated signal frequency and the clock bias is related to actual satellite frequency collected by the receiver, the actual satellite frequency can be defined as:

$$rf_i = f_i (1 + b) \tag{8.33}$$

Then, we could substitute Equations (8.32) into (8.33), like this:

$$f_i (1 + b) = tf_i \left(1 - \frac{(v_i - \dot{p}) \cdot l_i}{s}\right) \frac{s(f_i - tf_i)}{tf_i} + v_i l_i = \dot{p} l_i - \frac{s(f_i b)}{tf_i} \tag{8.34}$$

The next step is to expand the dot products $v_i l_i$ and $\dot{p} l_i$ in terms of the vector components, like this:

$$\frac{s(f_i - tf_i)}{tf_i} + v_{xi} l_{xi} + v_{yi} l_{yi} + v_{zi} l_{zi} = \dot{x} l_{xi} + \dot{y} l_{yi} + \dot{z} l_{zi} - \frac{s(f_i b)}{tf_i} \tag{8.35}$$

Then we can transform the set of equations (8.35) into a matrix form, like this:

$$\Delta p = \begin{bmatrix} \Delta pr_1 \\ \Delta pr_2 \\ \Delta pr_3 \\ \Delta pr_4 \end{bmatrix}, T = \begin{bmatrix} k_{x1} & k_{y1} & k_{z1} & 1 \\ k_{x2} & k_{y2} & k_{z2} & 1 \\ k_{x3} & k_{y3} & k_{z3} & 1 \\ k_{x4} & k_{y4} & k_{z4} & 1 \end{bmatrix}, \Delta q = \begin{bmatrix} \Delta x \\ \Delta y \\ \Delta z \\ -c\Delta b \end{bmatrix} \qquad (8.36)$$

By using the above matrices, we have:

$$\Delta p = T\Delta q \qquad (8.37)$$

As a result, we can find the adjustment Δq by using the following solution:

$$\Delta q = T^{-1}\Delta p \qquad (8.38)$$

Chapter Summary

The first part of this chapter has covered the traditional and modernized GPS signals. The second part has introduced the positioning algorithms used in GPS. In the next chapter, we look at the two different types of GPS, Differential GPS (DGPS) and Assisted GPS (AGPS) and implement some GPS applications using the iPhone.

Chapter 9

Differential GPS and Assisted GPS

I'm convinced that the only thing that kept me going was that I loved what I did. You've got to find what you love. And that is as true for your work as it is for your lovers. Your work is going to fill a large part of your life, and the only way to be truly satisfied is to do what you believe is great work.

Steve Jobs
2005

Chapter Contents

► Introducing different types of DGPS

► DGPS navigation message format

► What is AGPS?

► Implementing AGPS programs using iPhone

Differential Global Positioning System (DGPS) enhances the GPS by adding a local reference ground-based station to augment the information available from the satellites.

Since the ground-based stations are in the fixed locations, we know their precise position. The technique of DGPS is to use these stations to estimate and broadcast the difference between the positions calculated by the satellite systems and the known fixed positions of these stations. After calculating the difference, we can make use of this result to calibrate the errors among the measured satellite pseudoranges and the actual pseudoranges. Causes of errors can be multipath defractions, atmospheric distortion, satellite clock biases, and

Introduction to Wireless Localization: With iPhone SDK Examples, First Edition. Eddie C.L. Chan and George Baciu.
© 2012 John Wiley & Sons Singapore Pte. Ltd. Published 2012 by John Wiley & Sons Singapore Pte. Ltd.

receiver clock biases and orbital errors. These causes have been discussed in Chapter 7, Section 7.5.

DGPS runs in a long wave radio frequency in the band 285 kHz to 325 kHz. It is commonly used for land and air navigation, marine radio in waterways. Moreover, the ground-based stations have a shorter range of transmission than the satellite. Users should be near to the stations within 370 km.

9.1 Types of DGPS

In this section, we look at different types of DGPS. DGPS refers to any type of Ground Based Augmentation System (GBAS). There are many GBAS in the world, for example, European DGPS Network, United States National DGPS and Canadian DGPS. Many countries operate DGPS services based on the US National DGPS.

US National DGPS (NDGPS) is an enhancement of the previous Maritime Differential GPS (MDGPS). It started the expansion in the late 1980 and finished in March 1999. In the past, MDGPS only cover coastal area, the Great Lakes, and the Mississippi River inland waterways. But after finishing the NDGPS, the coverage area has been extended to the entire continental of US. Only few areas such as Rocky Mountain are in low or no coverage in US. The DGPS stations continuously broadcast the real-time DGPS signal to the receivers for correcting the estimation. Nowadays, there are 82 DGPS stations in NDGPS network, and it will increase to 128 stations in the next 15 years.

9.2 How DGPS Works

As mentioned, differential GPS involves the cooperation of many group-based stations which broadcast continuously the difference between the positions calculated by the satellite systems and their known fixed positions. There are two methods for calculating these difference: real-time DGPS and post-process DGPS.

9.2.1 Real-time DGPS

Real-time DGPS makes use of ground-based stations to broadcast the difference between the actual position and estimated position from satellites' signals. The difference can be broadcasted by either in the form of radio signal if the source is land-based or in the form of the satellite signal if it is satellite-based.

Let's have an example to illustrate how real-time DGPS works. In Figure 9.1, both the user receiver and the DGPS ground-based station receive the same satellite signal at the same time. The distance between user receiver and the station should not be over 370 km. Assume the actual position of the station be (x, y) and the station be the known position. After calculating from the satellite signal, the estimated position of the station is $(x + 10, y - 5)$, in other words, it means the difference is $(x + 10, y - 5)$. And the station broadcasts the difference to the receiver. The receiver calculates its current position to be $(x + 30, y + 40)$, while receiving the difference to be $(x + 10, y - 5)$ from the station. As a result, the user receiver uses the difference to adjust its position calculation to be $(x + 40, y + 35)$. The adjusted position usually is more accurate and more precise. The precision is around 3 to 5 meters in average.

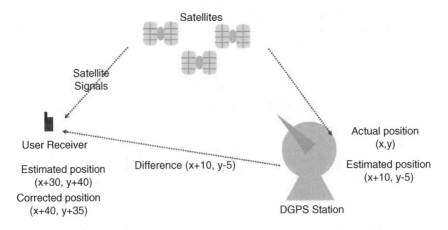

Figure 9.1   Example of real-time DGPS.

There are two navigation systems based on the real-time DGPS technique: Wide Area Augmentation System (WAAS) and Local Area Augmentation System (LAAS).

Wide Area Augmentation System (WAAS)

Wide Area Augmentation System (WAAS) is a precise and accurate navigator system over the continental United States, and portions of Mexico and Canada. WAAS was developed by the United States Department of Transportation (DOT) and the Federal Aviation Administration (FAA) in 1994. WAAS is not managed and maintained by the US military, it is controlled by FAA and DOT. WAAS is open to the public which can be used for civilian and commercial applications.

Before WAAS was developed, many positioning systems suffered poor accuracy due to errors, such as atmospheric errors, clock drift, and satellite position errors. Therefore, DOT and FFA developed WAAS to enhance the accuracy in aircraft navigation. The aim of WAAS is to provide a precise approach for aircraft landings and in-flight navigation. It includes altitude information and provides course guidance, the distance from the runway, and some elevation information at all points.

Instead of using the ground-based station to broadcast the difference data, WAAS uses its own geostationary satellites to broadcast signals. These satellites are at the fixed orbit over North America. For the ground segment, it is composed of number of Wide-area Reference Stations (WRS) and the Wide-area Master Stations (WMS). WRS is used for continuously monitoring and collecting the GPS signals. After collecting GPS signals, it sends to WMS using a terrestrial communication network.

When WMSs receive the data from WRSs, they will generate two sets of corrections: fast corrections and slow corrections. Fast corrections are designed to correct the rapid change errors, GPS satellite orbit errors and clock errors. The GPS receivers retrieve the fast correction data from the geostationary satellites, and then the receivers correct the position independently. Slow corrections are typically designed to correct the 'long-term'

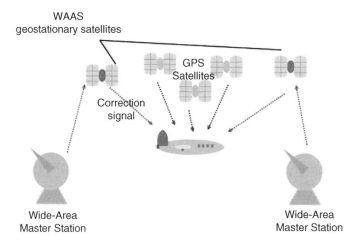

Figure 9.2   WAAS model.

errors, including ephemeris errors and ionospheric errors. Figure 9.2 shows how the WAAS model works.

In November 2006, there were 29 WRSs in WAAS, of which 20 of them were in the USA, seven in Alaska, one in Hawaii and one in Puerto Rico. And two WMSs were located at the Pacific and Atlantic coasts of the USA. To broadcast the corrections, WMSs need to transmit the corrections to their own geostationary satellites. The geostationary satellites broadcast the corrections using the basic GPS signal structure, which is in the L1 band on frequency 1575.42 MHz. All WAASs are available and compatible with GPS receivers. Receivers can retrieve the corrections from geostationary satellites. In addition, there are two more geostationary satellites which were named Galaxy XV and Anik F1R, and there are plans to increase up to three satellites in 2016.

Local Area Augmentation System (LAAS)

Local Area Augmentation System (LAAS) makes use of local reference receivers located around the destination position to improve the accuracy of GPS. LAAS is applied together with the WAAS in many major airports in the USA. LAAS provides a greater accuracy for the aircraft during takeoffs and landings. The reference GPS receivers near the runway give more accurate and precise corrections to the incoming or outgoing aircrafts.

Let's compare the performance between WAAS and LAAS, in terms of cost, scope and operating frequency. They share somehow similar techniques to improve the accuracy. First, the cost of LAAS is much lower than the WAAS. The scope of LAAS can only cover a specific local area, for example, a runway in an airport whereas WAAS can be deployed in much larger area, for example, cities or even continents. LAAS uses the shared band in radio frequency 108 MHz to 118 MHz which requires turning off some of the existing navigational aids in order to collect the information from LAAS. WAAS does not have this kind of problem which operates in a specific satellite frequency. As LAAS uses the shared band, the signal may be derogated due to the multipath problem and this leads to inaccurate positioning.

9.2.2   Post-process DGPS

Post-process DGPS is based on the position logs at known location to correct the position. The position logs are collected by base GPS receivers and rover GPS receivers which will be further processed by the office processing software, and then produces a correction file. Nowadays, there are many permanent GPS base stations which provide the logs data about GPS signals to process the correction files. These files are generally geo-tagged in the geographical information system (GIS) which data can be downloaded from the Internet or via a bulletin board system (BBS). The GPS base stations are consistent and run 24 hours continuously, so that the logs are reliable to fit with the real-time positioning process. This position logs can be sourced from public, commercial, web-based services and base station ownership.

Public Sources

Public sources are collected by the government unit. However, there are some legitimate concerns about the government data for public access, for example, legal liability and cost recovery. This will affect the government unit to have the decision of public access.

Commercial Sources

The base data are collected by the consulting companies and universities. Companies and universities provide the data through the Internet. Users can purchase the data at per day or per hour rates. It is the most expensive way to collect these raw logs data for post-processing.

Web-based Services

Web-based services provide a web-based platform for common users submitting the GPS data through the Internet, and then the web administrator will verify and upload the post-processed data to the GIS. It is the most economical way to process GPS data. Users do not require to know special techniques about the post-processing, and it saves time to train the users about post-processing.

Base Station Ownership

Base station ownership requires the user to own their base station which is responsible for collecting and processing the data. As the base station is owned by an individual unit, it is the most flexible way to run the post-process DGPS. However, the cost is high, since the owner should have their own equipment such as at least two GPS receivers, and the maintenance cost and setup fee are paid by the owner. If a large amount of data is required to be collected and processed, then the cost of the system will be more expensive.

9.3   DGPS Navigation Message Format

In Chapter 8, Section 8.1.4, we introduced the format of GPS message from the satellite. However, the format of GPS message is different from the DGPS message format. The standard format of DGPS message is developed by the Radio Technical Commission for Maritime Services (RTCM) Study Committer 104 (SC-104). RTCM was founded by the US

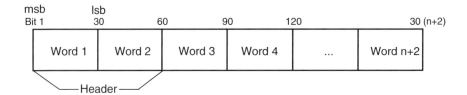

Figure 9.3   The format of the frame of RTCM SC-104 Version 2.3.

Government since 1947. And it becomes a nonprofit scientific and educational organization now. RTCM mainly focuses on the maritime radio-communication and radio-navigation which are all for the maritime applications. RTCM SC-104 messages are also applied in the GPS usage. In this section, we focus on the protocol of RTCM SC-104. Two versions of the protocols are popularly used nowadays. They are RTCM SC-104 version 2.3 and version 3.0.

9.3.1   RTCM SC-104 Version 2.3

RTCM SC-104 version 2.3 was released to the public in August 2001. Figure 9.3 shows the format frameworks of the version 2.3. The message contains a number of 30 bits of words. The first two words of the message frame is a header, and the rest of the words represent the navigation message. Figure 9.4 shows the format of the header of RTCM SC-104 Version 2.3.

Let's first introduce the DGPS message headers as depicted in Figure 9.4. The first word of the message contains preamble, frame ID, station ID, and parity. Preamble is a 8-bit variable, and its value is fixed to 0110 0110. Frame ID represents the type of message. The frame ID is a 6-bit parameter. There is a 10-bit parameter for station ID, and the ID represents the ID number of reference stations on the ground. After the station ID, there are 6 parity bits for checking error. Next is the second word in the message. The second word contains 5 parameters which are modified Z-count, sequence number, frame length, station health, and the parity bit. Modified Z-count represents the time reference for the message. It is a 13-bits parameter. Then, the 3-bit sequence number represents the number of the

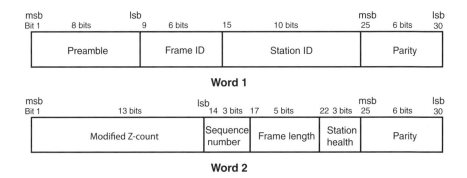

Figure 9.4   The format of the header of RTCM SC-104 Version 2.3.

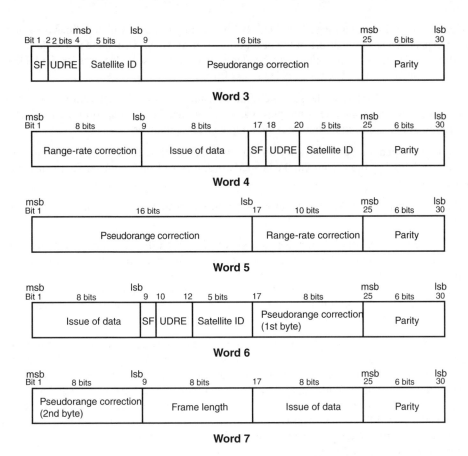

Figure 9.5   The format of the message type 1 of RTCM SC-104 Version 2.3.

current frame used, which counts in an ascending order. The function of sequence number is to monitor the frame synchronization. Thus, the frame length represents the length of next frame. Because the number of words in frame depends on the message type and number of the visible satellites, so the frame length is not fixed at each time. Finally, a 3-bit station health parameter represents the reference station whether it works properly or not.

There are 64 types which may display the frame ID. Message type 1 and 9 are the most important messages. We will focus on the message type 1 in following paragraph. Figure 9.5 shows the format of the message type 1 in version 2.3. There are some parameters which are important to DGPS message. They are scale factor, user differential range error, satellite ID, pseudorange correction, range-rate correction, and issue of data.

First, scale factor (SF) is a one bit flag, which represents the resolution of the pseudorange and range-rate corrections. When the flag is set, the resolution of the pesudorange and range-rate corrections is set to 0.32 meter and 0.032 meter per second. If the flag is unset, the resolution of the pesudorange and range-rate corrections become 0.02 meter and 0.002 meter per second.

Second, user differential range error (UDRE) is a 2-bit parameter which defines the range with expected one sigma errors of the pesudorange corrections. Also, there is a scale factor which refers to the station health in the header words. And the scale factor is for the UDRE to determine the range. The range is from $<= 0.1$ meter to > 8 meters.

Third, Satellite ID refers to the satellite number which has a DGPS correction. The satellite number for each satellite is unique. Because there are 32 satellites orbiting the Earth, therefore, 5 bits are used to represent the unique satellite number.

Fourth, the pseudorange correction calibrates the specific time for the target satellite. The time is provided by the Z-count in the header word. It is a 13-bit parameter for the pseudorange correction.

Fifth, the range-rate correction is an 8-bit parameter which is for the pseudorange rate correction.

Finally, Issue of Data (IOD) represents a set of navigation data for the corrections. The reference stations update the clock and ephemeris data in approximately every two hours. The changes of clock and ephemeris data with the IOD values are included in the message frame which refer to issue-of-data clock (IODC) and issue-of-data ephemeris (IODE). IODC is a 10-bit parameter and IODE is a 8-bit parameter. Combination of these two data helps to correct the ephemeris information.

Table 9.1 shows the message type in version 2.3. For message type 9, it is similar to message type 1. The format is the same as message type 1. Message type 1 allows every visible satellite to repeat the fields in the message; however, message type 9 only allows the number of satellites not greater than 3 in each message. It means that a stable clock in the reference station is important to the message type 9. Message type 18 to 21 are for carrier-phase DGPS. For the message type 18 and 19, they contain the reference station's uncorrected carrier phases and pseudoranges. Message type 20 and 21 contain the corrected carrier phases and pseudoranges.

9.3.2   RTCM SC-104 Version 3.0

RTCM SC-104 version 3.0 is the new protocol for DGPS message which was published on February 2004. The development of RTCM SC-104 version 3.0 is because the requirement of Real Time Kinematic (RTK) satellite navigation. Real Time Kinematic (RTK) satellite navigation is a carrier-phase enhancement of GPS used in land survey and in hydrographic survey. RTK uses the satellite's carrier as its signal, not the messages contained within the format. The format of version 3.0 is different from version 2.3, and it is a more efficient protocol than version 2.3.

Figure 9.6 shows the framework of RTCM SC-104 version 3.0. There is a variable length framework in version 3.0. And the components in the message include preamble, reserved, message length, data message, and the parity bits. First, the preamble is fixed to 8-bit parameter and the value is 11010011. It represents the starting point of the message. Second, there are 6 bits allocated for future use. And then following the reserved bits is the message length bits. The message length is 10-bit parameter for counting the length in the message. Next is the variable length of data message, which is from 0 bytes to 1023 bytes. And the last parameter is the 24 parity bits. The parity bits refer to cyclic redundancy checking.

Table 9.1   Message type of RTCM SC-104 V2.3.

Message type	Status	Details
0	–	Undefined
1	Fixed	DGPS corrections
2	Fixed	Delta DGPS corrections
3	Fixed	GPS reference station parameters
4	Tentative	Reference station datum
5	Fixed	GPS constellation health
6	Fixed	GPS null frame
7	Fixed	DGPS radiobeacon almanac
8	Tentative	Pseudolite almanac
9	Fixed	GPS partial correction set
10	Reserved	P code differential corrections
11	Reserved	C/A code L1, L2 delta corrections
12	Reserved	Pseudolite station parameters
13	Tentative	Ground transmitter parameters
14	Fixed	GPS time of week
15	Fixed	Ionosphere delay message
16	Fixed	GPS special message
17	Fixed	GPS ephemeredes
18	Fixed	RTK uncorrected carrier phases
19	Fixed	RTK uncorrected pseudoranges
20	Fixed	RTK carrier-phase corrections
21	Fixed	RTK/high-accuracy pseudorange corrections
22	Tentative	Extended reference station parameters
23	Tentative	Antenna type definition record
24	Tentative	Antenna reference point (APR)
25–26	–	Undefined
27	Tentative	Extended radiobeacon almanac
28–30	–	Undefined
31–36	Tentative	GLONASS messages
37	Tentative	GNSS system time offset
38–58	–	Undefined
59	Fixed	Proprietary message
60–63	Reserved	Multipurpose messages

For the message types in version 3.0, there are 13 message types in the primary release version. It supports GPS and Global Navigation Satellite System (GLONASS) in RTK applications. Table 9.2 is the list of the message types in version 3.0. There are four groups to summarize the 13 message types; they are observations, station coordinates, antenna

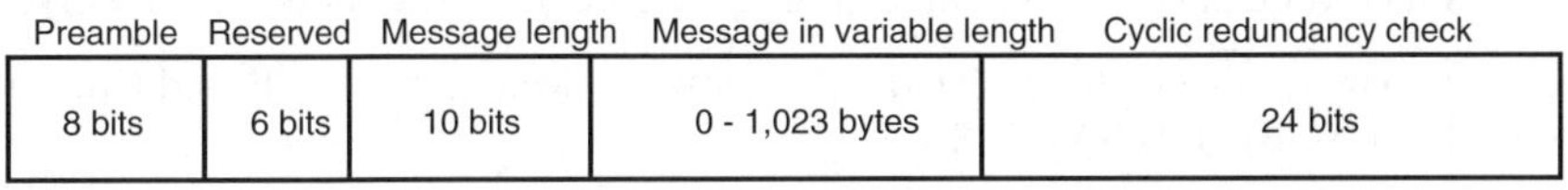

Preamble	Reserved	Message length	Message in variable length	Cyclic redundancy check
8 bits	6 bits	10 bits	0 - 1,023 bytes	24 bits

Figure 9.6   The format of the frame of RTCM SC-104 Version 3.0.

Table 9.2   Message type of initial release of RTCM SC-104 V3.0.

Message type	Status
1001	L1 only GPS RTK Observables
1002	Extended L1 only GPS RTK Observables
1003	L1 & L2 GPS RTK Observables
1004	Extended L1 & L2 GPS RTK Observables
1005	L1 only GLONASS RTK Observables
1006	Extended L1 only GLONASS RTK Observables
1007	L1 & L2 GLONASS RTK Observables
1008	Extended L1 & L2 GLONASS RTK Observables
1009	Stationary RTK reference station ARP
1010	Stationary RTK reference ARP with Antenna Height
1011	Antenna Descriptor
1012	Antenna Descriptor and serial number
1013	System Parameters

description, and auxiliary operation information. Moreover, it has a supplement message types added in version 3.0 in May 2006. They are GPS ephemeris, GLONASS ephemeris, network RTK (MAC), and proprietary messages.

9.4   Assisted GPS

The traditional GPS requires the receiver to search for the orbit and clock data of the satellite. This process is called 'Time To First Fix' (TTFF) or 'cold start.' It needs to take at least 30 seconds to finish these processes. If the users are located in the NLOS environment or highly interference area, it may even take several minutes. AGPS helps to address this problem.

AGPS makes use of the cellular network towers installed with GPS receivers to assist users to get the satellite information. First, GPS receivers at cellular network tower are always configured to collect the satellites' signals. When the user requests to locate his/her positioning, the cellular network tower will send those collected satellites' signal data to the user. Those data act as the 'real relevant data' from real satellites which are required to calculate their position.

There are several benefits of using AGPS. First, since the cellular signals can penetrate into buildings, AGPS can perform effective positioning tasks even in NLOS environment or somewhere that the signal is low in strength such as an urban area. Second, it obtains a reasonable precision between 5 meters to 50 meters. Third, AGPS reduces the TTFF time which is required to wait for identifying the relevant satellites, so now it can perform a faster location acquisition. Finally, since the number of satellite identification process is reduced, the power consumed in GPS receiver is also reduced. As a result, the battery life of the GPS receiver can now be sustained longer. Figure 9.7 shows how AGPS works.

There are two major modes for assisted GPS known as Mobile Station Based (MSB) and Mobile Station Assisted (MSA) GPS.

In MSB mode AGPS, the position is calculated by the receiver itself, and the MSB GPS receiver's job is to acquire the four types of assistance data: 1) time (date and time in UTC

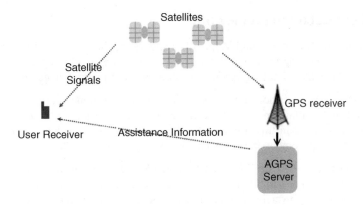

Figure 9.7   Assisted Global Positioning System.

standard), 2) reference frequency in mobile phone, 3) cell-tower positions and 4) almanac and/or ephermeris (to compute the relative satellite motion). The receiver will be responsible for directly computing the expected Doppler frequencies for all satellites. The offsets of relative satellite motion, receiver motion, and receiver oscillator are known as the expected Doppler frequencies. The receiver will receive this information to minimize the frequency search space.

In MSA mode AGPS, the position is calculated at a server and the MSA GPS receiver's job is to acquire the reference time, reference frequency and expected satellite Doppler rate data. With this data, the AGPS device receives signals from the visible satellites and sends the measurements to the AGPS server. Then AGPS server calculates the position and sends it back to the AGPS device.

9.5   AGPS in iPhone

iPhone includes three types of positioning technologies: Global Positioning System (GPS), Cellular Positioning System (CPS) and Wi-Fi Positioning System (WPS). Let's look at the performance of each positioning system in iPhone.

GPS in iPhone takes advantage of the cellular signals to assist and enhance the GPS features, such that users can require less satellite information for calculating the position. In iPhone, GPS and CPS are stuck together and form the AGPS. However, CPS only get a very rough position estimation. The error distance is generally around 20 to 50 m depending on the density of cell towers distributed in an area.

WPS in iPhone requires to grasp the information about Wi-Fi signal strengths and access points from different locations. Since it is still not feasible to have a database which stores these Wi-Fi information of every building in the world, it can only achieve a few tens of meters' precision. So that in Chapter 4, we introduced the implementation of a customized and better Wi-Fi positioning system.

In comparing the performance of WPS in iPhone, AGPS in iPhone is a much more mature and accurate technology for positioning. For these reasons, AGPS usually dominates in usage.

9.5.1 Core Location Framework

The Core Location framework and MapKit framework are the software libraries for developing AGPS and map application. If you do not have the frameworks in your Mac machine, please go to iPhone Dev Center (http://developer.apple.com/iphone/) to get the frameworks.

We now briefly introduce the Core Location framework in iPhone. Core Location Framework is used for localization in iPhone. There are three positioning systems (mentioned before) which are used in this framework. The users determine which technique is used according to the requirement and environment.

If we're going to use the Core Location framework, please be aware of the usage of the power. iPhone consumes a lot of energy to use this framework to locate itself continuously. The consumption of power in iPhone highly depends on the precision level. The more precision required, then the more power is consumed.

Core Location Manager

Inside the Core Location framework, there is a main class CLLocationManager, which is in charge of configuring and communicating with the Core Location. The CLLocationManager class defines the interface for configuring the delivery of location-and heading-related events to the application.

To begin the Location Manager, we need to create a CLLocationManager instance of this class, for example:

```
CLLocationManager *locationManager = [[CLLocationManager alloc]
          init];
```

Now, we can further establish the parameters that determine when location and heading events should be delivered and to start and stop the actual delivery of those events.

The next step is to create a Location Manager delegate for itself. The delegate obeys CLLocationManagerDelegate protocol. And the delegate is used to receive location and heading updates from a CLLocationManager object. The Location Manager uses the delegate to receive the positioning information, and it may need to take a few seconds. Here is the code of creating the delegate and assigning to itself:

```
LocationManager.delegate = self;
```

After creating the delegate, we still need to set two configurations in Location Manager. They are the desired accuracy and distance filter.

Setting the Desired Accuracy

Desired accuracy means the users request a degree of accuracy in localization. Before starting the positioning, users should determine what precision of the positioning is required. For example, if the users would like to know which city they are located, then the degree of accuracy could be set lower. The users should notice when they request more accuracy, then more time and more power in iPhone are used. Moreover, the desired accuracy does

Table 9.3 Types of desired accuracy in Location Manager.

CLLocationAccuracy types	Desired accuracy
kCLLocationAccuracyBest	-
kCLLocationAccuracyNearestTenMeters	10 meters
kCLLocationAccuracyHundredMeters	100 meters
kCLLocationAccuracyKilometer	1 kilometer
kCLLocationAccuracyThreeKilometers	3 kilometers

not guarantee the actual accuracy is the same as the requested accuracy. The following is an example of setting the desired accuracy.

```
locationManage.desiredAccuracy = kCCLocationAccuracyBest;
```

desiredAccuracy is the property in the Location Manager. The accuracy uses the CLLocationAccuracy value, and the type of the CLLocationAccuracy is double. There are several options of CLLocationAccuracy. First, kCLLocationAccuracyBest is the default value of desiredAccuracy, which is the most accurate for the estimated location. Second, kCLLocationAccuracyNearestTenMeters is that the user determines the accuracy which is within ten meters. Third, kCLLocationAccuracyHundredMeters is to set the accuracy within a hundred meters. Fourth, kCLLocationAccuracyKilometer means that the accuracy is within a kilometer. Finally, kCLLocationAccuracyThreeKilometers is the least accuracy in the settings which represents the desired precision as being within three kilometers. Table 9.3 shows the options of the CLLocationAccuracy.

Setting Distance Filter

The delegate of Location Manager keeps updating the changes of the user's position. If the user keeps updating in every change, there may have been many minor changes in the location when the user is moving slowly. As a result, the resources are wasted in updating the minor changes. The distance filter helps to solve the problem when the users would like to reduce updating the minor changes of the location. It can reduce to display some zigzag shape of walking path when the users move slowly. Users could set the distance filter in the Location Manager. The distance filter will not update the minor position changes until a significant position change is made which is more than the setting filter value. For example, the user sets the distance filter as ten meters, then the Location Manager will not update the position until it is ten meters further away from the previous position. The following is the example of the setting of the distance filter.

```
locationManager.distanceFilter = 10.0f;
```

distanceFilter is the property in the Location Manager. The accuracy uses the CLLocationDistance function, and the CLLocationDistance shown in meters. The above example means that the distanceFilter is set to be ten meters. If the user would like to have no filter, the value of distanceFilter can be set to kCLDistanceFilterNone:

```
locationManager.distanceFilter = kCLDistanceFilterNone;
```

Starting the Location Manager

After setting the Location Manager, we need to check whether the location service in iPhone has been enabled or not. If the location service has not been enabled, then the Location Manager cannot work although it starts. There is a Boolean property locationServicesEnabled for checking the location services. The following is an example of checking the location service in the Location Manager.

```
if ([locationManager locationServicesEnabled]){
    // start location manager
}
```

When we tell the Location Manager to start, it will start to use the delegate method to estimate the current location. The delegate method will keep updating the user's location which exceeds the distance filter. The following is the example that starts updating the location.

```
[locationManager startUpdatingLocation];
```

Stopping the Location Manager

There is a method to stop the Location Manager to keep updating the changes of the location. The method is called stopUpdatingLocation; it will tell the Location Manager to stop the delegate for updating the location. When users do not need to update the location, stopUpdatingLocation method stops updating the location and it can also disable the appropriate hardware. The advantage of stopping the updating location method is to save power and resources in iPhone. When the users would like to estimate the current location again, then they can use startUpdatingLocation to restart the updating location method again. The following is the example that stops updating the location.

```
[locationManager stopUpdatingLocation];
```

Location Manager Delegate

The CLLocationManagerDelegate protocol defines the methods to receive location and heading updates from a CLLocationManager object. The methods of the protocol are optional. There are two methods which the delegate in Location Manager for responding to location events, which are for updating the location and handling errors. The methods are called locationManager:didUpdateToLocation:fromLocation: method and locationManager:didFailWithError: method.

Updating the Location

When the delegate in the Location Manager would like to estimate the current location, it calls the locationManager:didUpdateToLocation:fromLocation: method. There are three parameters in the method. The first parameter means using the method of the CLLocationManager object. The second parameter is the CLLocation object which represents the current location information. The third parameter is also the CLLocation object which represents the previous updated location. When the method is started, the third parameter is nil because there is no previous update. The following is an example of the locationManager: method.

```
-  (void)locationManager:(CLLocationManager *)manager
      didUpdateToLocation:(CLLocation *)newLocation
              fromLocation:(CLLocation *)oldLocation {
          // update the location
          // manage the new location and old location
}
```

Error Occurred in Delegate Protocol

The second option method is locationManager:didFailWithError:. When the delegate is unable to retrieve a location value, then locationManager:didFailWithError: method is used. There are two parameters in the method. The first parameter means that the CLLocation-Manager object is unable to retrieve the location. The second parameter is the error object containing the reason why the location could not be retrieved.

There are four types of error that the delegate protocol causes error. When the error occurs, the method will return the error code. They are kCLErrorLocationUnknown, kCLErrorDenied, kCLErrorHeadingFailure and kCLErrorNetwork.

kCLErrorLocationUnknown error means the Location Manager cannot retrieve the right location. And the Location Manager keeps trying until it works. When the users get this error message, the simple solution ignores the message, and waits for the new event of localization.

kCLErrorDenied error means the application cannot access the Location Manager. In other words, the location service is not available to the application. When the user receives this error message, the user should stop the location service and restart it.

kCLErrorHeadingFailure error occurs when the heading cannot determine the right estimation. It may be caused by a strong interference from nearby magnetic fields. The user can try the application at the other location.

kCLErrorNetwork error occurs when the network is unavailable or a network error occur. When the user uses the WPS technique for localization, if the network is not unavailable, then the error will occur. The user should check whether the network service has been enabled or not.

The following is the example of the locationManager:didFailWithError: method.

```
-  (void)locationManager:(CLLocationManager *)manager
        didFailWithError:(NSError *)error {
      // display the error message
}
```

9.5.2   Core Location

When the location information is collected by the delegate in Location Manager, the Location Manager uses instances of the CLLocation class to store the location information. There are five location attributes for storing location information. They are coordinate, altitude, horizontalAccuracy, verticalAccuracy, and timestamp.

Coordinate shows the geographical coordinate information. It is an instance of CLLocation-Coordinate2D class. The coordinate includes two variables, they are latitude and longitude.

For the latitude, positive values indicate latitudes north of the equator and negative values indicate latitudes south of the equator. For the longitude, it is a zero at meridian and it has positive values extending east of the meridian and negative values extending west of the meridian. Both latitude and longitude are in degrees in measurement. The following is an example of getting coordinate in Location Manager.

```
CLLocationDegrees latitude = newLocation.coordinate.latitude;
CLLocationDegrees longitude = newLocation.coordinate.longitude;
NSString *coordinate = [[NSString alloc] initWithFormat:@"<\%g◀¢X,
        \%g◀¢X>", latitude, longitude];
```

Altitude indicates meters above sea level. If the current location is below sea level, then the altitude is a negative value. Altitude refers to the CLLocationDistance function, and the CLLocationDistance returns double. The following is an example of getting the altitude.

```
CLLocationDistance altitude = newLocation.altitude;
```

horizontalAccuracy is an attribute in CLLocation which means the radius of uncertainty for the location. It measures in meters. The higher value of horizontal accuracy means a lower level of confidence in obtaining accuracy of location. On the contrary, the lower value of horizontal accuracy represents the high confidence of accuracy of estimation. horizontalAccuracy refers to the CLLocationAccuracy function, and the CLLocationAccuracy returns double value in meters. The following is the example of getting the horizontalAccuracy.

```
CLLocationAccuracy hAccuracy = newLocation.horizontalAccuracy;
```

verticalAccuracy is an attribute in CLLocation which represents the confidence of the estimation of altitude. Similar to the horizontalAccuracy, the higher value of vertical accuracy means a lower level of confidence in obtaining the correct altitude. verticalAccuracy also refers to the CLLocationAccuracy function, and the CLLocationAccuracy returns double value in meters. If the returned value is negative, it means that the altitude value is invalid. The following is the example of getting the verticalAccuracy.

```
CLLocationAccuracy vAccuracy = newLocation.verticalAccuracy;
```

Timestamp records the time when the location is estimated. It is stored in NSDate class.

Measuring the Distance between Coordinates

The CLLocation is not only stored in the above five attributes. There is a method of calculating the distance between two coordinates. The method is called getDistanceFrom: method. It inputs two CLLocation objects to the method. The first CLLocation object is fromLocation, which is the starting coordinate for the measurement. The second CLLocation object is toLocation, which is the destination of the coordinate. The method returns in meters. The following is an example of getting the distance between two coordinates.

```
CLLocationDistance distance = [fromLocation getDistanceFrom:
        toLocation];
```

Getting Speed and Course Information

CLLocation class has two more attributes to store the movement information of the iPhone. They are speed and course information.

Speed is a property in CLLocation class, which represents the speed of the device movement per second. It measures in meters. If the speed is a negative value, it means that the speed in invalid. speed refers to the CLLocationSpeed function, and the CLLocationSpeed returns double value in meters. The following is an example of getting the speed of the device movement.

```
CLLocationSpeed speed = newLocation.speed;
```

Course shows the direction of the device movement. It measures in degree, which starts at due north and continues clockwise around the compass. In other words, it returns 0 degree when the direction is north, 180 degrees when the direction is south. If the course returns a negative value, it means that the direction cannot be determined. The following is an example of the course.

```
CLLocationDirection direction = newLocation.course;
```

9.5.3   GPS Program in iPhone

After this briefing on the Core Location Framework, let's get started on the GPS program in iPhone. This program shows the location information collected by Location Manager. Here is code of the header file of the localization program:

```
// GPSMonitorViewController.h program example\newline

#import $<$UIKit/UIKit.h$>$

#import $<$CoreLocation/CoreLocation.h$>$ \newline

@interface GPSMonitorViewController : UIViewController $<
        $CLLocationManagerDelegate$>${
    IBOutlet UILabel *latOutput;
    IBOutlet UILabel *longOutput;
    IBOutlet UILabel *altOutput;
    IBOutlet UILabel *haccOutput;
    IBOutlet UILabel *vaccOutput;
    IBOutlet UILabel *directionOutput;
    IBOutlet UILabel *currenttimeOutput;
    IBOutlet UILabel *distanceOutput;
    IBOutlet UILabel *startlocOutput;
    IBOutlet UIButton *resetBtn;
    CLLocation *startingPoint;
    CLLocationManager *locationManager;
}

@property (nonatomic,retain) IBOutlet UILabel *latOutput;
@property (nonatomic,retain) IBOutlet UILabel *longOutput;
@property (nonatomic,retain) IBOutlet UILabel *altOutput;
@property (nonatomic,retain) IBOutlet UILabel *haccOutput;
@property (nonatomic,retain) IBOutlet UILabel *vaccOutput;
@property (nonatomic,retain) IBOutlet UILabel *directionOutput;
@property (nonatomic,retain) IBOutlet UILabel *currenttimeOutput;
@property (nonatomic,retain) IBOutlet UILabel *distanceOutput;
```

```
@property (nonatomic,retain) IBOutlet UILabel *startlocOutput;
@property (nonatomic,retain) IBOutlet UIButton *resetBtn;

@property (nonatomic, retain) CLLocation *startingPoint;
@property (nonatomic, retain) CLLocationManager *locationManager;

- (void)locationManager:(CLLocationManager *)manager
    didUpdateToLocation:(CLLocation *)newLocation
           fromLocation:(CLLocation *)oldLocation;

- (void)locationManager:(CLLocationManager *)manager
       didFailWithError:(NSError *)error;

-(IBAction)resetStartLocation;

@end
```

We have created the header file for the GPSmonitor program. Let's discuss what we need to notice in the header file. First, we need to use the Core Location framework, so we import the header of the Core Location framework. Because the Core Location framework is not a general framework such as UIKit framework, the user should type the import header files by themselves.

Second, we conform this class to the CLLocationManagerDelegate method, so that we can collect the location information from the Location Manager.

Third, we create the CLLocationManager object for holding the instance of the Core Location. And we create the CLLocation object to store the first updating location information. The CLLocation object is for calculating the distance between current location and the starting location. Furthermore, we declare a number of labels for the information display on the application. After we declare the objects, we set the property of each objects.

After that, we define the methods which we will use in the GPSMonitorViewController.m file. In the application, there are three methods: updating the location, handling the error message, and the method in the reset button, which is for calculating the distance between current location and the starting location.

After defining the objects and methods in the header file, we move on to create the main file. Because the source code is a bit long, we break down the code into a number of methods to explain and help users understand the workflow of the application easily. First, when the application starts, the method -(void)viewDidLoad is loaded. As shown below, the following source code is about the viewDidLoad method, we initialize and allocate the CLLocationManager instance, and set the delegate method. Moreover, we set the configurations of Location Manager. They include desired accuracy and distance filter. When we are ready to start the Location Manager, we run the startUpdatingLocation method to start updating the location information.

```
...
- (void)viewDidLoad{
    [super viewDidLoad];
    self.locationManager = [[CLLocationManager alloc] init];
```

```
        locationManager.delegate = self;
        locationManager.desiredAccuracy = kCLLocationAccuracyBest; //
                best accuracy
        locationManager.distanceFilter = kCLDistanceFilterNone; // no
                filter
        [locationManager startUpdatingLocation];
}
```

If errors occured in the Location Manager, the locationManager:didFailWithError: method is used. Otherwise, the locationManager:didUpdateToLocation:fromLocation: method is run for updating the current location. In this method, it receives the updating location information and stores at newLocation object, and there is previous updated location information at oldLocation object. We display the location information on the view such as coordinate, altitude, timestamp, and we also set the format to display the number and date through NSNumberFormatter and NSDateFormatter. The following is the part of code of updating location information.

```
...
- (void)locationManager:(CLLocationManager *)manager
     didUpdateToLocation:(CLLocation *)newLocation
            fromLocation:(CLLocation *)oldLocation{

    static NSNumberFormatter *numberFormatter = nil;
    if (numberFormatter == nil){
        // set the number format on display
    }

    static NSDateFormatter *dateFormatter = nil;
    if (dateFormatter == nil){
        // set the date format on display
    }

    // display the location information
...
}
```

Moreover, there is a reset button for resetting the starting location. It is used for calculating the distance between the current location and the starting location. There is a simple method called resetStartLocation. We get the current location at the Location Manager and store the startingPoint object. Therefore, the distance between starting-Point and newLocation is calculated during updating current location in the locationManager:didUpdateToLocation:fromLocation: method.

```
...
-(IBAction)resetStartLocation{
    self.startingPoint = locationManager.location;
}
```

Figure 9.8   The result of GPS monitor application program.

Figure 9.8 shows the result of our first GPS application program. It displays the current location information, and the distance between the current and starting locations. Because the application runs in the simulator, the current location is at the fixed values.

9.5.4   Core Location Heading

In the Core Location framework, there are two different ways to get the direction information. The first way is in the CLLocation class. Users would get the speed and the direction of the movement device through speed and course which are the property in the class. They are very frequently used in tracking device movement in the navigation application. However, the direction information provided by the CLLocation class is rough and is not precise enough. Therefore, we introduce the second way to measure precise direction information in the Core Location framework.

The iPhone device contains a magnetometer which can provide more precise direction information, and it includes instances of the CLHeading class. The advantage of using the magnetometer is that it is more precise, and the user does not need to keep moving with the device. It is always used in the compass-based applications. In this section, we briefly introduce the CLHeading class.

What is Magnetometer?

A magnetometer is a device which measures the magnetic fields originating from the Earth. By using the measurement result, it could calculate the precise orientation of the device. However, the magnetometer sometimes may be affected by local magnetic fields which originated from other electronic devices. As a result, the precision of estimation becomes worse. Core Location framework can filter out much interference from the other electronic devices, and maintain the high degree of precision in the estimation of heading.

Starting the Location Heading

To detect the heading direction, we first need to check whether the hardware magnetometer in iPhone has been enabled or not. If it has not been enabled, then the location heading cannot work although it starts. There is a Boolean property headingAvailable for checking whether the magnetometer is available or not. The following is an example of checking the device for the heading measurement.

```
...
if (locationManager.headingAvailable){
      // start location manager
}
```

When we tell the Location Manager to start updating the heading, it will start to use the delegate method to measure the magnetic fields. The delegate method will keep updating the heading information. The following starts updating the heading information.

```
[locationManager startUpdatingHeading];
```

If users get enough heading information, they may want to stop updating. There is a method called stopUpdatingHeading, which will tell the Location Manager to stop the delegate from updating the heading information. The magnetometer device also stops by using this method, which helps to save power. The following stops updating the heading information.

```
[locationManager stopUpdatingHeading];
```

Updating the Location Heading

When the delegate in the Location Manager gets the updating heading information, it calls for the locationManager:didUpdateHeading: method. There are two parameters in the method. The first parameter means using the method of the CLLocationManager object. The second parameter is the CLHeading object which represents the current heading information. The following is an example of the locationManager:didUpdateHeading: method.

```
...
- (void)locationManager:(CLLocationManager *)manager
        didUpdateHeading:(CLHeading *)newHeading{
          // update the current heading information
          // manage the current heading information
}
```

Core Location Heading Attribute

In the CLHeading object, heading information is contained which is generated by a CLLo-cationManager object. There are four attributes in CLHeading object. They are headingAc-curacy, magneticHeading, trueHeading, and timestamp.

headingAccuracy means the maximum deviation between the reported heading and the true geomagnetic heading. It predicts accuracy of the heading information in magneticHeading property. If the headingAccuracy value is high, it means that the degree of accuracy is low. On the other hand, if the headingAccuracy value is low, it indicates that the accuracy is high. However, if the headingAccuracy value returns a negative value, it means that the heading information is invalid. It may be caused by the interference from other local magnetic fields. headingAccuracy refers to the CLLocationDirection function, and the CLLocationDirection returns double. The value is measured in degrees. The following is an example of obtaining heading accuracy.

```
CLLocationDirection headingAccuracy = newLocation.headingAccuracy;
```

magneticHeading is one of the important attributes in the CLHeading class. It represents the heading that points toward the magnetic North Pole. Therefore, when the magneticHeading value is 0, it means that the heading points toward the magnetic North Pole. 90 represents magnetic East, 180 represents magnetic South, and 270 represents magnetic West. How-ever, when the headingAccuracy property is a negative value, it means that the value of magneticHeading becomes unreliable. magneticHeading also refers to the CLLocationDi-rection function. And it also measures in degrees. The following is an example of getting the degree of magnetic heading.

```
CLLocationDirection magneticHeading = newLocation.magneticHeading;
```

trueHeading is related to true North. The heading points toward the geographic North Pole. It is different from magneticHeading property which points toward the magnetic North Pole. When trueHeading value is 0, it means that the heading points towards to geographic North Pole, and 180 represents geographic South, and so on. If the trueHeading property returns a negative value, it represents the heading is invalid and it could not be determined. trueHeading also depends on the CLLocationDirection function. And it also measures in degrees. The following is an example of getting the degree of magnetic heading.

```
CLLocationDirection trueHeading = newLocation.trueHeading;
```

Timestamp records the time when the location is estimated. It stores the NSDate class.

Accessing the Raw Heading Data

There are three properties to represent a raw geomagnetic data: x, y, and z properties.

x property represents the x-axis deviation from the magnetic field lines being tracked by the device. The range of value is -128 to +128.

y property represents the y-axis deviation from the magnetic field lines being tracked by the device. The range of value is -128 to +128.

z property represents the z-axis deviation from the magnetic field lines being tracked by the device. The range of value is -128 to +128.

9.5.5 Compass in iPhone

After introducing CLHeading class, we show a simple program to help users to understand how to apply the CLHeading class in an iPhone application. In this application, we use the CLHeading class to create a compass and include the coordinate information. Here is the code of the header file of the compass application:

```
// CompassViewController.h program example

#import $<$UIKit/UIKit.h$>$
#import $<$CoreLocation/CoreLocation.h$>$
#import $<$CoreLocation/CLLocationManagerDelegate.h$>$

@interface CompassViewController : UIViewController $<
      $CLLocationManagerDelegate$>${
    CLLocationManager *locationManager;
    IBOutlet UIImageView *compassImage;
    IBOutlet UILabel *locOutput;
}

@property (nonatomic, retain) UIImageView *compassImage;
@property (nonatomic, retain) CLLocationManager *locationManager;
@property (nonatomic, retain) IBOutlet UILabel *locOutput;

- (void)locationManager:(CLLocationManager *)manager
      didUpdateToLocation:(CLLocation *)newLocation
            fromLocation:(CLLocation *)oldLocation;

- (void)locationManager:(CLLocationManager *)manager
         didFailWithError:(NSError *)error;

- (void)locationManager:(CLLocationManager *)manager
         didUpdateHeading:(CLHeading *)newHeading;

- (void) rotate: (double) degrees;

@end
```

We have created the header file for the Compass program. First, just like the previous GPS program, we use the Core Location framework, so we need to #import the header of Core Location framework <CoreLocation/CoreLocation.h>. Thus, we conform this class to the CLLocationManagerDelegate method, so that we can collect the location information and heading information from the Location Manager.

After that, we create locationManager as the CLLocationManager object for holding the instance of the Core Location. And we declare compassImage as UIImageView, which is used for the compass image. And locOutput is a label which stores the current position coordinate. After declaring the objects, we set the property of each object.

The next step is to define the methods which we will use in the controller file. In the application, there are four methods used in application. They are updating the location, handling the error message, updating the heading in Location Manager, and the method for rotating the compass image.

After defining the objects and methods in the header file, we move on to create the main file. To start the application, it runs the - (void)viewDidLoad method. First, we initialize and allocate the CLLocationManager instance, and set the delegate method. Then we set the configurations of Location Manager. It includes desired accuracy, distance filter, and heading filter. After we are ready to start the Location Manager, we run the startUpdatingLocation and startUpdatingHeading method to start updating the location information.

We can now focus on the updating heading information method in this application. We use locationManager:didUpdateHeading: method to update heading information. The method is simpler than the updating location information. In this application, we use the magnetic heading for the compass. After we get the magnetic heading, and then it passes the value to the rotate method which rotates the compass image. The following is the method for updating heading information.

```
...
- (void)locationManager:(CLLocationManager *)manager
        didUpdateHeading:(CLHeading *)newHeading{
    double magneticHeading = newHeading.magneticHeading;
    [self rotate: magneticHeading];
}
```

For the image rotation, there is a function CGAffineTransformMakeRotation() for image transformation. CGAffineTransformMakeRotation(angle) is in counterclockwise rotation when the angle is positive. Therefore, when we use the magnetic heading as an angle input, it will rotate automatically. The following is the method for image rotation.

```
...
-(void) rotate:(double) magneticHeading{
    [UIView beginAnimations:nil context:NULL];
    compassImage.transform = CGAffineTransformMakeRotation(
            magneticHeading);
    [UIView commitAnimations];
}
```

Figure 9.9 shows the result of compass application program. When the Location Manager receives the updating location information and heading information, it displays the coordinate on locOutput label and rotates compassImage in the degree of magnetic heading.

9.5.6   MapKit framework

In the previous section, we discussed the Core Location framework in the iPhone. Core Location framework receives location information. However, the location information is in terms of a group of numbers, so that users may get confused and not know where they are. Therefore, we should convert the location information into the graphic user interface to help users to comprehend the current location. The MapKit framework embeds the map interface in the application. It contains a lot of classes in handling and managing the location information in the map interface. Figure 9.10 shows the output of the program.

Figure 9.9   The Result of Compass Application Program.

By using MapKit framework, we can provide a fully functional map into the application. The map supports many features such as displaying street-level map image, zooming and panning the map, and supporting the touch events. We implemented a *GPSinGoogleMapView-Controller* class to demonstrate these features in the following program. We first define the *GPSinGoogleMapViewController* class, implementing *CLLocationManagerDelegate* in the header with properties like mapView, mapSlider for zooming, locationManager for positioning the current user as the code below:

```
#import <UIKit/UIKit.h>
#import <CoreLocation/CoreLocation.h>
#import <MapKit/MapKit.h>

@interface GPSinGoogleMapViewController : UIViewController <
       CLLocationManagerDelegate> {
    IBOutlet MKMapView *mapView;
    IBOutlet UISlider *mapSlider;
    IBOutlet UILabel *locationOutput;
    CLLocationManager *locationManager;
}
```

Figure 9.10   MapKit.

```
@property (nonatomic,strong) IBOutlet MKMapView *mapView;
@property (nonatomic,strong) IBOutlet UISlider *mapSlider;
@property (nonatomic,strong) IBOutlet UILabel *locationOutput;
@property (nonatomic, strong) CLLocationManager *locationManager;

-(IBAction)zoom;
-(IBAction)backLocation;

- (void)locationManager:(CLLocationManager *)manager
     didUpdateToLocation:(CLLocation *)newLocation
            fromLocation:(CLLocation *)oldLocation;

- (void)locationManager:(CLLocationManager *)manager
        didFailWithError:(NSError *)error;

@end
```

In the implementation of *GPSinGoogleMapViewController* class, we synthesize the properties at the beginning, and initialize the locationManager by assigning the delegate object to itself. We also assign the desired accuracy and distance filter for the locationManager.

After finishing the initialization of locationManager, we can start it and the updated positioning information will keep being sent to the delegation methods.

```
#import "GPSinGoogleMapViewController.h"

@implementation GPSinGoogleMapViewController

@synthesize mapView;
@synthesize mapSlider;
@synthesize locationOutput;
@synthesize locationManager;

- (void)viewDidLoad{
    [super viewDidLoad];
    self.locationManager = [[CLLocationManager alloc] init];
    locationManager.delegate = self;
    locationManager.desiredAccuracy = kCLLocationAccuracyBest; //
            Best accuracy
    locationManager.distanceFilter = kCLDistanceFilterNone; // no
            filter
    [locationManager startUpdatingLocation];
}
```

There are two delegation methods, *didUpdateToLocation* and *didFailWithError*, for the CLLocationManagerDelegate protocol; they will be invoked when positioning information is available and errors found during positioning respectively. In the *didUpdateToLocation* method, we show the current user location coordinates at the locationOutput label. In the *didFailWithError* method, we just skip it for simplicity.

```
- (void)locationManager:(CLLocationManager *)manager
        didUpdateToLocation:(CLLocation *)newLocation
                fromLocation:(CLLocation *)oldLocation{
    NSString *locationstring = [[NSString alloc] initWithFormat:@"
            <\%g^\circ, \%g^\circ>", newLocation.coordinate.
            latitude, newLocation.coordinate.longitude];
    locationOutput.text = locationstring;
}

- (void)locationManager:(CLLocationManager *)manager
        didFailWithError:(NSError *)error{
}
```

For the *backLocation* method, we show the current user location in the map and center the map view with current user location also by using the *setCenterCoordinate* method of MapView with animation. To handle the zooming event driven by the map slider, the current value of map slider will be obtained and used in zooming the map accordingly.

```
-(IBAction)backLocation{
    mapView.showsUserLocation = TRUE;
    CLLocation *location = [locationManager location];
    CLLocationCoordinate2D coordinate = [location coordinate];
```

```
    CLLocationCoordinate2D centerlocation = coordinate;
    [mapView setCenterCoordinate:centerlocation animated:YES];
}

-(IBAction)zoom{
    MKCoordinateRegion region;
    MKCoordinateSpan span;
    span.latitudeDelta=125*(1-mapSlider.value)+0.01;
    span.longitudeDelta=0.001;
    region.span=span;
    region.center=mapView.centerCoordinate;
    [mapView setRegion:region];
}

@end
```

Chapter Summary

This chapter has introduced different types of DGPS and AGPS. There are two types of DGPS: Real-time DGPS and Post-process DGPS. Real-time DGPS makes use of ground-based stations to broadcast the difference between the actual position and estimated position from satellites' signals while post-process DGPS is based on the received position logs at a known location to calibrate the position. There are two major modes to assisted GPS known as Mobile Station Based (MSB) and Mobile Station Assisted (MSA) GPS. This chapter has also covered the iPhone implementation about AGPS, compass and the map. In the next chapter, we cover other types of existing positioning technologies, such as acoustic-based, vision-based and RFID-based.

Chapter 10

Other Existing Positioning Systems

We are very careful about what features we add because we can't take them away.

Steve Jobs
1955–2011

Chapter Contents

▶ What is energy consumption model?

▶ What is acoustic localization and its methodologies?

▶ What is vision-based localization and its methodologies?

▶ What is RFID technology and its components?

The positioning systems we've learnt so far in this book are about Wi-Fi and GPS. In this chapter, we're going to get a whole lot more with other positioning technologies. In fact, other positioning technologies also commonly use similar methods we've discussed in Chapter 3, for example, angle of arrival, time difference of arrival, triangulation and etc.

So this chapter focuses mainly on how they perform in theories and their applications. We will look at three types of positioning technologies: acoustic-based, light-based and RF-based tracking. We could say sound, light and RF are all governed by most of physics wave theories and algorithms. They will lose energy when they are traveling in space. Most of these positioning technologies make use of the loss of energy when they travel in space to estimate the location. Before we look at each of the positioning technologies in detail, we first look at the energy consumption model.

Introduction to Wireless Localization: With iPhone SDK Examples, First Edition. Eddie C.L. Chan and George Baciu.
© 2012 John Wiley & Sons Singapore Pte. Ltd. Published 2012 by John Wiley & Sons Singapore Pte. Ltd.

Energy Consumption Model

Techniques for acoustic-based, light-based and RF-based localization have relied exclusively upon the energy consumption model. Usually, this model involves sensors to emit, sense, transmit and receive the wave (light, sound or RF).

Assume there are k sensors deployed in a n by m field grid; the activity begins with a target source (unknown position) to emit the wave and reach these K sensors; there are three basic energy consumptions in this activity: broadcasting, sensing, and receiving. Total energy consumption, denoted as E, is:

$$E = E_b + E_s + E_r \tag{10.1}$$

where E_b, E_s and E_r are the energy consumed in broadcasting, sensing, and receiving respectively.

As energy may also be lost in the environment, the energy consumed in broadcasting decreases exponentially with the traveling distance:

$$E_{b_i} = E_b \cdot \|x - x_i\|^{-\alpha} + n_i \tag{10.2}$$

where E_{b_i} is the energy left in broadcasting after traveling $|x - x_i|$ distance, x is the position of the source (target), x_i is the known position of the ith sensor, α is the path loss and n is the white Gaussian noise with zero mean and variance σ^2. As we can measure E_{b_i}, α and n, we can use Equation (10.2) to find the position of the source.

10.1 Acoustic-based Positioning

Acoustic-based positioning (or acoustic localization) makes use of sound waves to determine the distance and direction of an unknown object. There are two basic types of acoustic positioning: active and passive. Figure 10.1 shows what the active and passive acoustic localizations are.

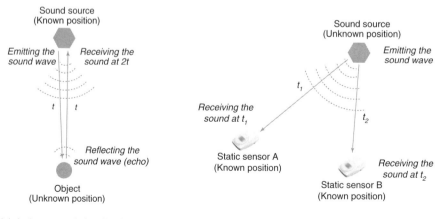

(a) Active acoustic localization   (b) Passive acoustic localization

Figure 10.1   Two types of acoustic localization.

10.1.1 Active Acoustic Positioning

Active acoustic positioning actively creates a sound source (known position) when the sound pulse hits the target object (unknown position) and it will reflect to produce an echo. By measuring the time difference from emitting the sound and receiving the echo, we can estimate the distance and direction of the target object. Figure 10.1(a) shows the case of active acoustic localization and t represents the signal traveling and returning time. The algorithms used in this approach are similar to the RTOF algorithm in Chapter 3, Section 3.2.5.

10.1.2 Passive Acoustic Positioning

Passive acoustic localization does the localization the other way round. Static sensors (known positions) are placed to receive the sound from a sound source (unknown position). A sound source will generally take a different amount of time to reach each sensor. By measuring the time of arrival (or equivalently, time delay) of the signal for each sensor, the location of the sound source can be determined. Figure 10.1(b) shows the case of passive acoustic localization and t_1 and t_2 represent the traveling time to the sensor A and B respectively. We can similarly apply algorithms of time of arrival (TOA) and time difference of arrival (TDOA) to this approach. Algorithms of TOA and TDOA are covered in Chapter 3, Section 3.2.2 and Section 3.2.4.

10.1.3 What is Beamforming?

Beamforming is a signal processing technique which can operate in both transmission and reception and creates signals, which at particular angles experience constructive interference, while others experience destructive interference. The use of constructive interference (and removing noise with destructive interference) improves the signal transmission/reception.

For example, if we simply send a sharp sound pulse to underwater to detect the position of submarine, it will fail because the ship will first hear the echo pulse that happens to be nearest the ship, then hear later echo pulses that happen to be further from the ship. This makes positioning difficult.

The beamforming technique is applied in this situation. The sonar equipped in the ship involves sending the pulse at slightly different times or different frequencies (the sonar projector closest to the ship last), so that when every pulse hits the submarine, every pulse reflects and produces the echo pulse to hit the ship at exactly the same time and the ship receives the effect of a single strong echo pulse because of constructive interference.

There are two types of beamformers: conventional and adaptive.

Conventional Beamformers

Conventional beamformers makes use of fixed time-delays or fixed angles to emit wave pulses and strong echo wave pulses will come back with the combination of all the reflected waves. One of the applications using this conventional beamformer is the Green Bank telescope in the National Radio Astronomy Observatory in West Virginia, USA. It helps to provide a 100 m clear aperture.

Adaptive Beamformers

Adaptive beamforming techniques usually operate in the frequency domain which create fixed frequency-delayed wave to induce constructive interferences such that strong echo wave pulses come back. The destructive interferences also help to improve rejection of unwanted signals from other directions. This beamformer is applied to military submarine which detects enemies by sonar.

10.1.4   Applications of Acoustic Positioning

In land and aerial robotics, the positioning can be achieved by incorporating Global Positioning System (GPS) measurements. The GPS signal is not available in an underwater environment and other techniques to correct the drift errors of the measurement units. Acoustic localization operates perfectly in underwater to detect fish, submarines and autonomous underwater vehicles. However, this localization approach performs less effectively in the non-line-of-sight (NLOS) environment, such as heavily-built cities and crowded areas.

10.2   Vision-based Positioning

Vision-based positioning (or light-based positioning) makes use of different visual scenes captured by one or multiple sensors to estimate the position. This positioning approach usually relies on optical sensors rather than ultrasound, dead-reckoning and inertial sensors.

Starting from 2009, Google has been sending its vehicles to take pictures of the streets in many cities and these pictures were uploaded and tagged with the position in Google Maps. Users can check pictures of the landmarks to see whether they go in the right places. In fact, these activities illustrate the example of the vision-based positioning in a manual manner.

Basically, there are two types of vision-based positioning: camera-based or landmark-based.

10.2.1   Camera-based Positioning

Camera-based positioning makes use of a camera or optical sensor to take pictures from a reference object and measures the pixel differences of distorted pictures to estimate the position.

For example, assume that we do not know the fixed position of the target object. An optical sensor is moved to take different pictures to the target object in different positions. The focal length, f and changing positions of the optical sensor (or camera) are known and marked down. If the optical sensor takes picture far away from the target object, the target object looks smaller proportionally. If the optical sensor takes picture at angle θ, the target object projected into picture will be distorted at angle θ. These pixel differences are used to estimate the position of the target object.

Figure 10.2 shows the perspective projection in the camera-based positioning model. The (x_r, y_r, z_r) is a 3D coordinate of the real world object and (x_p, y_p) is a 2D coordinate of the image. The real world object is projected onto the image plane. The relationship for this perspective projection is:

$$x_p = f \cdot \frac{x_r}{z_r}, \quad y_p = f \cdot \frac{y_r}{z_r} \qquad (10.3)$$

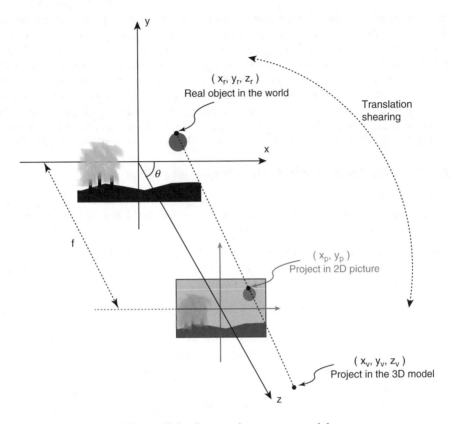

Figure 10.2   Perspective camera model.

The angle θ can be obtained if the focal length f is known or simply measuring the facing direction of the optical sensor manually. θ and f are the distortion factors which control the rates of translation and shearing of the pictures. This problem is transformed to be a classic computer graphics problem and can now be handled by matrix operations:

$$\overline{X_p} = \overline{RX_r} + \overline{T} \tag{10.4}$$

where $\overline{X_p}$ is the object project plane matrix, $\overline{X_r}$ is the real world plane matrix, $\overline{R}$ and $\overline{T}$ are the rotation and translation matrices.

We can obtain the (x_p, y_p) easily by measuring the picture. Now we have three unknown values (x_r, y_r, z_r) and we need to take four different pictures from different perspectives. After we get the set of equations, we can calculate the position using linearization (discussed in Chapter 8, Section 8.3.4) or the hyperbolic method (discussed in Section 3.2.4).

Problems of Camera-based Positioning

The first problem of this positioning system is that it is difficult to match the features of the objects in the real world and the objects in one or more pictures, so that we can reference on some feature points to determine the position and orientation of the sensor.

In order to do this, we need to find multiple features in one or more pictures. For example, in Figure 10.2, the circle (sun) is a feature object that can be recognized and matched. A feature point of the circle is used to form the projection matrix. The second problem is that the multiple object recognitions and matchings are computational intensive processes. These operate fairly slowly to estimate the location. The third problem is that adopting this approach in a large and complex environment is unimplementable. It is not possible to take and store pictures of all places in a huge area, such as campus sites and cities. The environment may also be changed frequently for example, the movement of the users and the changing position of the furniture. These make the position estimation not possible.

10.2.2   Landmark-based Positioning

Landmark-based positioning is somehow an optimized version of camera-based positioning which only matches at simple features such as points and lines, complex patterns or characterized objects to estimate the position.

After we take several pictures of the target object in an unknown position, we first need to do feature extractions of all the pictures such that we can identify points and lines, complex patterns or characterized objects as special features.

We then make use of the special features to form a set of equations similarly in Equation (10.4) to calculate the position using the hyperbolic (discussed in Chapter 3, Section 3.2.4) and linearization approaches (discussed in in Chapter 8, Section 8.3.4). Figure 10.3 summarizes the general steps for performing the landmark-based positioning.

Feature extractions, recognitions and matchings are the fundamental techniques in image analysis and recognition domain. Current landmark recognition techniques can be mainly categorized into two groups: 1) template-based representation uses holistic texture features,

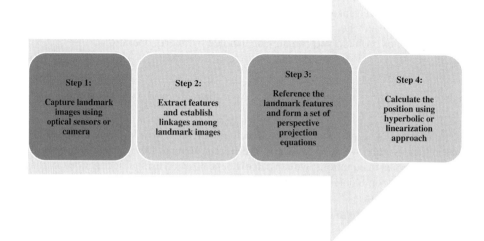

Figure 10.3   Steps for landmark-based positioning.

stores the whole building/location image pattern in an array and compares them using a suitable metric such as the Euclidean distance; 2) feature-based representation uses geometrical features of the landmark, for example by extracting the relative position and attributes of distinctive features of pillars, crossbeams, walls, doors and windows and internal decorations. We briefly talk about these two approaches.

Template-based Approach

One of the most widely accepted algorithms in the template-based approach is the Eigenspace Method. This method is based on Principal Components Analysis (PCA) technique. One of its main advantages is the dimensionality reduction scheme that enables landmark recognition to be performed rapidly.

Fisher-space is another well-known template-based approach. It is based on Linear Discrimination Analysis (LDA). LDA is a classical statistical technique using the projection which maximizes the ratio of scatter among the data of different classes to the scatter within the data of the same class. Features obtained by LDA are useful for pattern classification since they make data of the same class closer to each other and the data of different classes further away from each other.

Experiments show, however, that the eigen-space and fisher-space methods are not robust in dealing with variations in lighting conditions. To overcome these problems, wavelets decomposition is adopted to break images into approximations and details of different levels of scales and work on several approximations of each landmark image. It is generally applied together with a common feature-based approach, Gabor wavelet.

Feature-based Approach

The feature-based approach extracts landmark features from the major components of the location such as pillars, crossbeams, walls, doors, windows and internal decorations.

The Gabor wavelet filter can be applied for the enhancement of the eigen-space and fisher-space respectively which provides robustness against varying brightness and contrast of landmark images. This filter captures the properties of spatial localization, orientation selectivity, spatial frequency selectivity and quadrature phase relationship seems to be a good approximation to the filter response profiles encountered experimentally in a building. Gabor wavelets are used for landmark analysis because of their geometrical relevance and computational properties. They have been found to be particularly suitable for image decomposition and representation when the goal is the derivation of local and discriminating features. Wavelet decomposition has been widely used in various applications such as palmprint recognition, object recognition and Chinese and English language character recognition.

The Gabor wavelets (kernels, filters) can be defined as follows:

$$\Psi(k, x) = \frac{k^2}{\sigma^2} \exp\left(-\frac{k^2 x^2}{2\sigma^2}\right) \left[\exp(ik.x) - \exp\left(\frac{\sigma^2}{2}\right)\right] \qquad (10.5)$$

The multiplicative factor k^2 ensures that filters tuned to different spatial frequency bands have approximately equal energies. The term $\exp(-\sigma^2/2)$ is subtracted to render the filters

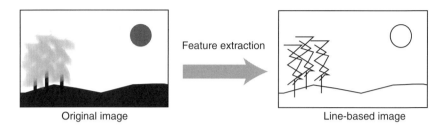

Figure 10.4   Feature extraction.

insensitive to the overall level of illumination. The Gabor wavelet representation allows the description of spatial frequency structure in the image while preserving information about spatial relations. The complex amplitude of the transforms is used as features to test for the presence of spatial structure and restricted to a band of orientations and spatial frequencies within the Gaussian envelope. The amplitude information degrades gracefully with shifts in the image location at which it is sampled over the spatial scale of the envelope.

$$O_{k,x} = I \otimes \psi(k, x) \tag{10.6}$$

The convolution outputs (both the real part and the magnitude) of a sample image exhibit strong characteristics of spatial locality, scale and orientation selectivity corresponding to those displayed by the Gabor wavelets. Such characteristics produce salient landmark features, such as pillars, crossbeams that are suitable for forming the perspective projection matrix. Figure 10.4 shows an example of feature extractions in which the original image is converted to a line-based image.

After we have found the feature points, we again form a set of equations about the projection to calculate the position. The other parts of the calculation are similar to camera-based positioning.

10.2.3   Applications of Vision-based Positioning

One of the major applications of vision-based positioning is the robot navigation. The primary goal of robot navigation is to guide a mobile robot to a desired destination. Inside the memory of the robot pictures of landmarks and obstacles are stored in the pre-specified route. In order to locate itself, the robot is equipped with one or multiple optical sensors (or cameras) to capture visual scenes in the surrounding environment. Vision-based positioning techniques are applied to measure the difference between visual scenes and estimate the position of the robot.

10.3   What is RFID Technology and Its Components?

Radio Frequency Identification (RFID) technology makes use of tags and readers to track the objects. For example, an item is attached to a tag and the tag emits the RF wave. The signal strength of RF wave decreases exponentially with its traveling distance. A RFID reader (transceiver or antenna) is used to detect and receive the RF waves that come out from the RFID tag. The reader is connected to a PC or embedded into a mobile device to estimate the position of the object using the energy loss equation defined in Equation (10.2).

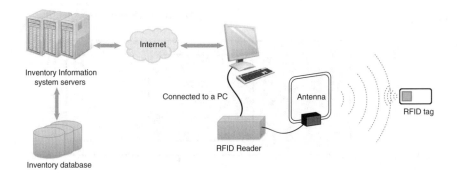

Figure 10.5   RFID system.

The PC or mobile device also stores the identification information transferred from the tag. Figure 10.5 shows a typical RFID system.

With this basic idea of how the RFID system operates, this section further introduces the basic components of RFID system which consists of a tag and a reader. Then, we cover how each component tracks items and finally talk about applications of the RFID system.

10.3.1   RFID Reader

An RFID reader continuously creates an encoded radio signal and transmits to search for one or more tags. The RFID also supplies tags with energy by emitting its RF waves and orders tags to respond with the identification (ID) information. There are two types of RFID reader: static and mobile. A static reader generally connects to a PC via the communication lines, while a mobile reader is a handheld device which transfers the ID information of RFID tags via a Wi-Fi network.

In industrial practice, multiple RFID antennas are connected to a RFID reader to increase the chance to search for the tag, since the orientation of the tag is not guaranteed (a tag may be faced in an opposite direction to the reader). An RFID antenna is usually very big and is placed near a conveyer belt portal where it can conduct successful readings to the tags easily.

There are two ways to communicate between readers and tags: inductive coupling and radiating RF waves. In the first way, the antenna coil of the reader induces a magnetic field in the antenna coil of the tag. The tag then uses the induced field energy to communicate data back to the reader. For this reason inductive coupling only applies in a relatively short distance communication (less than half meter). In the second way, the reader radiates the energy in the form of RF waves (usually operates with a passive tag). Some portion of the energy is absorbed by the tag to power up the tags circuit. After the tag wakes up, some of the energy is reflected back to the reader. The reflected energy can be modulated to transfer the data contained in the tag.

10.3.2   RFID Tag

An RFID tag or transponder is an identification device attached to the item to be tracked. Tags are usually located on the containers which carry multiple items. Tags come in all

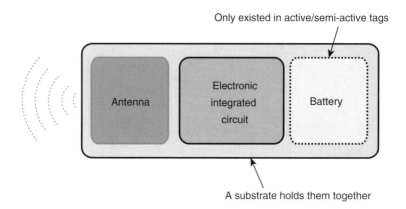

Figure 10.6   Components of RFID tag.

sorts of shapes, sizes, and capabilities. There are many different types of RFID tag with varying features and functions, but the common primary goal of the tag is to transfer the ID information of the tracked item. Different tags have different storing ID information capabilities. There are three general types: 1) read only, 2) write once read many (WORM), and 3) read-write.

An RFID tag is a microchip (electronic integrated circuit (IC)) combined with an antenna and a holder to attach the IC and the antenna together to an item. The IC is used to store and process information by modulating and demodulating a radio-frequency (RF) signal. The antenna is used to receive and transmit the RF wave. Figure 10.6 shows the components of a RFID tag.

RFID tags can be either passive, active or semi-passive depending on how they are powered. In the following paragraphs, we further describe the feature and function of each type.

Passive RFID Tag

Passive RFID tag does not have an internal power source (battery). The antenna of a passive tag receives the electromagnetic (EM) wave generated from the reader. The tag generates its own supply voltage by rectifying the induced voltage from the reader's EM wave. The design of the passive tag is much simpler and less expensive, as it does not have limited shelf life and has no battery. So this leads to a substantial amount of commercialization and applications with this type of RFID tag.

The major advantages of all passive tags are their lower production cost, smaller size, greater operational life and the fact that they are more robust in many environmental conditions. Most RFID scientists foresee that passive tags are the future of RFID. As long as the price of passive tags drops gradually, an increasingly larger deployment of passive tags is becoming possible.

The disadvantages of passive RFID tags are their limited detectable range and the fact that they have less memory to be utilized. Current technology can only allow 6 meters (20 feet) detectable range where the reader must be very close to the passive tag to transmit the signal. Generally, the passive tag only has 16 to 64 Kbits memory to store the data.

Active RFID Tag

In contrast to passive RFID tags, active RFID tags have an integrated power source which usually use small batteries to supply the circuitry and to generate the response data. As the active tag requires an extra battery, the size of the active tag is larger and the price is more expensive.

The major advantages of all active tags are their larger detectable range and better identification capability. The range of active tags is very often far superior to that of passive tags. They broadcast high frequencies from 850 to 950 MHz that can be read 30 meters (100 feet) or even farther away over 100 meters (300 feet) if larger capacity batteries are embedded into active tags. In contrast to the memory of passive tag, active tags can allow more user available memory (128 kb in general but can be more) and GPS capabilities can be built on top of it. Therefore, the tags can transmit not only the identification information but also the current position. With the GPS technology, the physical location of a cargo container can be traceable if an active tag with GPS feature is attached to the cargo container.

The disadvantages of active RFID tags are larger size, higher production cost, less operational life and continuous maintenance. Since active tags contain batteries, the size will be larger and the production cost will be higher. Electronic batteries such as alkaline, nickel-cadmium, nickel metal hydride have a life of around 3 to 5 years. They need to be detached from active tags after the batteries run down. Battery replacement can be very troublesome if the tag is mounted in an accessible position. Other alternatives are to use rechargeable batteries such as lithium chemistry, but they are significantly more costly.

Semi-active RFID Tag

The semi-active RFID tag combines the beauty and functionality of both active and passive tags. The semi-active RFID tag has an integrated battery but only supports the internal power usage of the tag itself whereas the active tag uses its internal battery power to transmit and receive signals from its antenna. And the semi-active tags can generate their own supply voltage by rectifying the induced voltage from the reader's RF wave. In this case, the battery life is much longer than the active one.

10.3.3   RFID Positioning

RFID positioning generally characterizes and adopts the indoor localization techniques and algorithms (covered in Chapter 3). The characteristics of RFID positioning are a kind of 'centralized' positioning system, since RFID tags have very limited capabilities in term of transmitted range and energy which only enable the short-range positioning. RFID positioning can be classified into two types: propagation-based and location-fingerprinting-based (LF-based).

Propagation-based RFID Positioning

Propagation-based RFID positioning estimates the position by measuring the received signal strength from adjustable long-range active RFID tags with path loss. Multiple readers collect signal strength measurements from an unknown position of a tag and estimate its position using triangulation methods, such as angle of arrival (AOA), time of arrival (TOA), and time

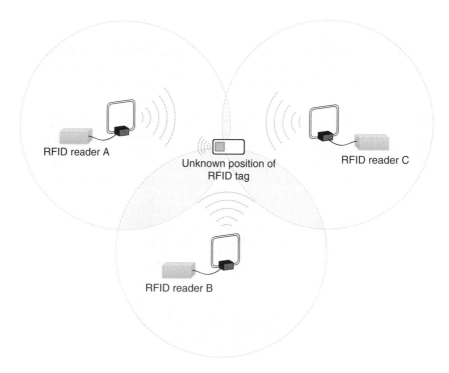

Figure 10.7   RFID triangulation positioning method.

difference of arrival (TDOA). We have covered these methods in Chapter 3, Section 3.2. Figure 10.7 shows a typical RFID triangulation method setup.

LF-based RFID Positioning

LF-based RFID positioning makes use of a set of reference fixed RFID tags with known positions on the covered area and collects the RSS measurements for a training database. Figure 10.8 shows the LF-based RFID positioning setup. We can estimate an unknown position of an active tag by by matching the similarity of RSS measurements in the LF training database. The basic principle and algorithms of this positioning type are covered in Chapter 3, Section 3.3.

10.3.4   Applications of RFID Positioning

RFID positioning can be applied together with the GPS that will allow for covering both short-range and long-range tracking of goods in the supply chain. RFID plus GPS technology enable visibility for both traders and customers to have access to inventory, orders, raw materials, and delivery points at any place, at any moment. This improves coordination among entities in the supply chain, for example, efficient delivery scheduling, accurate estimation of inventory ordering, optimally allocating labors to handle the transportation of inventory and etc.

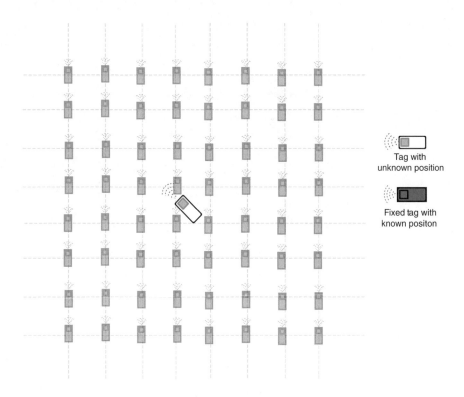

Figure 10.8   LF-based RFID positioning setup.

Chapter Summary

This chapter has covered other three types of positioning technologies: acoustic-based, vision-based and RFID-based. Other positioning systems such as acoustic-based and light-based are most effective in relatively open and flat outdoor environments but are much less effective in non-line-of-sight (NLOS) environments such as hilly, mountainous or built-up areas. For example, most acoustic localization applications not only require the sound source to have a high intensity and to be continuously propagated, they are also limited to localizing only within the area covered by the sound. Acoustic-based positioning generally operates underwater to detect fish, submarines and autonomous underwater vehicles.

Vision-based positioning usually operates in the robot navigation which makes use of different visual scenes captured by one or multiple sensors equipped in the robot to estimate its position. RFID-based positioning is generally applied in supply chain management systems which track the delivery of tagged products.

Part III
Applications in Wireless Localization

Chapter 11

AI for Location-aware Applications

I'll always stay connected with Apple. I hope that throughout my life I'll sort of have the thread of my life and the thread of Apple weave in and out of each other, like a tapestry. There may be a few years when I'm not there, but I'll always come back.

Steve Jobs
Feb. 1985

Chapter Contents

► What is the location-aware application?

► What are the AI techniques used in location-aware applications?

► Implementing the Tourist Guide application

O ver the last several years, many commercial and government organizations as well as university campuses have deployed WLANs such as IEEE 802.11b. This has fostered a growing interest in location-based services and applications. Apple's iPhone employs a multi-touch interface and brings a full degree of mobility. One of the most challenging problems in retrieving the location-aware information is to understand the behavior of users and fit it into the current location. In this chapter, the background of AI techniques and location-aware applications are covered. It also covers the implementation of a location-aware application in iPhone which has features of positioning, providing reviews, setting reminder alarms, providing suggestions for restaurants and guiding the user to the destination.

Introduction to Wireless Localization: With iPhone SDK Examples, First Edition. Eddie C.L. Chan and George Baciu.
© 2012 John Wiley & Sons Singapore Pte. Ltd. Published 2012 by John Wiley & Sons Singapore Pte. Ltd.

11.1   What is Location-aware Application?

'Location-awareness' means having certain levels of 'intelligence' to sense and respond to location-based information and events. 'Location-awareness' can be implemented to mobile devices like iPhone or other web-enabled phones that can passively or actively determine their location. iPhone provides AGPS to identify where users are if they choose to share their location-based information. Location-aware applications deliver online content to users based on their physical location.

In 2011, Google promotes Google+ to allow users to communicate to friends nearby and send out a request for video hangout (conferencing) to meet for coffee or dinner. While such applications provide increased social connectivity, they also create a highly targeted marketing opportunity for retailers and enhanced environmental awareness, offering users a location-based filter for online information.

Many location-aware applications have been developed recently such as suggestions for dining places, notifications about nearby traffic accidents and congestions, a nap alarm to remind the user to attend meetings or notices about a critical drop of a stock.

How Does It Work?

Location-aware application consists of three parts: positioning, location-based information and AI techniques. Figure 11.1 shows three components of the location-aware application.

Positioning can be facilitated by using GPS satellites, cell towers, Wi-Fi access points, or a conjunction of all these methods. Previous chapters in this book covered most of the existing positioning methods. Each method has trade-offs, unavailable time and environment. It is always more reliable to use one or more methods.

Location-based information can be divided into two types: static and dynamic. Static location-based information can be used to represent data that are fixed and do not change

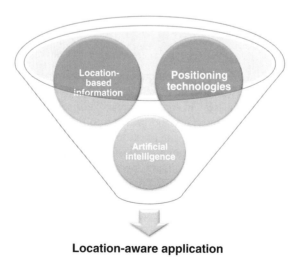

Location-aware application

Figure 11.1   Three components of the location-aware application.

over time, such as floor plans, street maps, building facilities' data etc. Dynamic location-based information usually involves one or more providers giving, generating, modifying the data on demand and is needed to update frequently, such as weather information (provider: observatory departments), traffic information (provider: reporting system driven by users and traffic departments) and the availability of services (service providers such as restaurants, cinema etc.).

Artificial intelligence techniques are very often applied to match and answer user queries and provide accurate information where users access location-aware (e.g., pervasive computing-enabled) applications and services. The location-aware information is then organized and presented in an efficient way according to users' preferences. Not only that, but AI should be able to give suggestions or even proactively plan for what users want to have.

What are the Implications for Industries and Academia?

Location-aware applications are significant and crucial for the commercial activities. Location-aware content is furnished for the convenience of the trading, the provision of advertising, marketing and merchandising services. Customers can use any service or product advertised and found on the web and find out the location of actual services nearby. The expectations of the customers' experience now include the geotagging of blogs and websites. Dealers can simply tag their replies to their customers on the embedded geolocation information in photos, e-mail, or Twitter posts.

Location-aware device can help students and researchers to tag photos, research data or other information in Google Maps. When the geotagged photos, videos, and audio tracks are available in the public domain, educators may establish effective windows into other study. Usage of the location-aware technology provides a very good opportunity for the experiential study. Geo-tagging supports field research by embedding location-aware applications which can effectively connect geo-tagged resources to researchers whenever and wherever they need. There are many useful location-specific metadata (coordinated or place names) including pictures, videos, blogs, or websites. By using location-aware browsers on mobile phones, users on nature walks not only can know the name of plants but also see community-generated pictures of related specific specimen.

Geo-tagging could be turned into educational games that use location-aware mobile devices; for example, in augmented reality simulations, students can use given clues and riddles to locate members of study groups at the site. Security officials can also help by giving advice to student in monitor trips across campus to let them finish the game safely. Useful data will be collected that contribute toward identifying where there are greater security risks, and show where facilities improvements are required. Librarians could base their work on readers' habits and queries and point them to locations where important books and media are placed. To conclude, the location-aware technology gives another platform for teachers to connect their teaching to geographic locations.

Concerns about Privacy and Security

There are two major concerns about location-aware applications: privacy and security. As interconnectivity between social networking sites is increasing like Twitter, Facebook, and Flickr, data posted on one site could link and show to another site without the endorsement of author(s) and users may not be aware that this has happened.

Spreading fault location-aware information is another major problem. For example, someone may raise a false flag of terrorist attacks in a location. If it is not handled properly, this could create public threats and chaos. Privacy and security controls should be carefully designed to allow different levels of sharing and security. Furthermore, indicators to the location-aware information should be suggested to identify the trust level of the source information.

What Is the Future of Location-aware Applications?

Many more mobile devices are used by more people in more locations. Location-aware becomes more and more effective in helping us predict things happening in the geographical space around us because location-aware applications offer larger amounts of data about the environment. These applications also offer knowledge of the physical world that help the supply of information become more convenient. Users can even add their own comments by using applications such as Facebook, posting notes on the nearest geolocation wall.

Now, technologists have started to think how location-aware devices lead to a new generation of 'citizen environmentalists' who can report location-specific information around them by using cell phones. Similar to the rise of citizen journalism, citizen environmentalism helps scientists to understand environmental health and quality depend on the specific contributions of informal data collection. Citizens can get information on air quality in weather forecasts and traffic conditions by using smartphones or tablet PC devices.

11.2   What are AI Techniques?

Artificial intelligence is the process of creating machines that can act in a manner that could be considered by people to be intelligent. The central problems of AI include such traits as reasoning, knowledge, planning, learning, communication, perception and the ability to move and manipulate objects.

In October 2011, Apple's iPhone 4S featured the AI personal assistant app called Siri bringing natural language processing to the smartphone. Initially it will support English, French and German with more languages and more features being added over time, according to Apple. The intention was to create an automated assistant that has some 'intelligence' to learn from the user and handle a variety of tasks. The project was lead by SRI International, a California based research institute; when it came to an end a number of people working on the CALO project founded Siri.

Figure 11.2 shows the daily uses of Siri. Siri can help you access most services just by saying it and replies with the right information. You can perform tasks such as:

► Finding your position

► Sending emails and text message without typing

► Setting alarms to remind meetings

► Reporting the weather forecast.

- Action: 'Tell my wife I'm running late.'
  Response: Text message is sent to your wife.

- Action: 'Remind me to call the vet.'
  Response: A reminder is set based on location, time or date

- Action: 'Any good burger joints around here?'
  Response: Your location is determined and a nearby restaurant is listed.

Figure 11.2 Examples of conversation with Siri in iOS 5.

Technologists are also beginning to consider the possibilities for these mobile assistants to inspire a new generation of smartphone devices. Users can order cell phones to do things just by talking the way you talk. Moreover, these mobile assistants are proactive and will plan for the next move of the user and give suggestions.

There is a huge possibility that future mobile applications will incorporate different kinds of intelligence by using AI techniques stemming from:

► fuzzy logic,

► natural language processing,

► data mining,

► intelligent agents,

► neural networks and

► semantic techniques etc.

In the following paragraphs, we briefly talk about two major AI techniques frequently used in location-aware applications: Fuzzy Logic and Natural Language Processing.

11.2.1 Fuzzy Logic

Usually, people are sensible to some abstract concepts, such as far away, near, cheap or expensive. The real world rarely operates in this fashion; many conditions can be partially

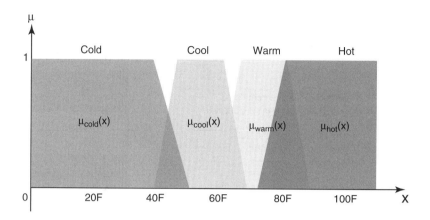

Figure 11.3   Human temperature perception.

true or partially false (or both). Fuzzy logic is a form of many-valued logic; it deals with reasoning that is approximate rather than fixed and exact. Fuzzy logic was proposed by Prof. Lotfi Zadeh in 1963.

In software, fuzzy logic was introduced to allow software to operate in the domain of degrees of truth. In contrast with traditional logic theory, where binary sets have two-valued logic – true or false – fuzzy logic variables may have a truth value that ranges in degree between 0 and 1. Fuzzy logic has been extended to handle the concept of partial truth, where the truth value may range between completely true and completely false. Furthermore, when linguistic variables are used, these degrees may be managed by specific functions.

Fuzzy membership function represents the degree of truth as an extension of valuation which provides a generalization of the indicator function in fuzzy sets. A fuzzy set X is any set that allows its members to have different degrees of membership, called membership function $\mu(x)$ in the real unit interval [0, 1]. Membership functions can either be chosen by the user arbitrarily, based on people's experiences and perspectives or machine learning methods. There are different shapes of membership functions: triangular, trapezoidal, piecewise-linear, Gaussian, bell-shaped, etc.

A fuzzy membership graph is plotted to show each fuzzy interval. Figure 11.3 shows an example of fuzzy membership graph which uses fuzzy logic to represent the human temperature perception (cold, cool, warm and hot).

Fuzzy logic in a location-aware application usually deals with many conditions when choosing a desired location, e.g. restaurant and hotel. Some conditions are chosen from a finite set of options, known as discrete choice, for example, restaurant type and location, which is easy to select from a set of multiple options. Some conditions are presented in range of values for selection, known as continuous choice, for examples, price and food rating. Different people will have their own perspective on value and flavor of food, which makes decision-making vague.

Using fuzzy logic, the continuous choice will transform into discrete choice, using the semantic term, by means of fuzzy membership function, which encapsulates the partial truths of the semantic term among the underlying continuous value. Thus, decision-making becomes direct and natural by means of semantic terms by using fuzzy logic, and this logic can get rid of numeric values.

Later in this chapter, we will apply fuzzy logic to implement a Tourist Guide application to represent the distance and price between the restaurant and the user location.

11.2.2 Natural Language Processing

In the 1950s, natural language processing (NLP) was introduced to handle the interaction between computers and human (natural) languages. Natural language processing is usually implemented by machine learning by which computer machines learn human languages from successive training processes. In the training process, a computer machine is usually exposed to a huge number of sample data. Algorithms or rules will be recognized and induced by the machine training methods, such as decision tree, association rule learning, genetic programming, artificial neural networks, support vector machines, clustering analysis and etc.

An NLP system in the location-aware application involves in understanding questions from the user, determining answers, organizing the answer with sufficient content to reply the user. For example, a user can order the NLP system like this, 'I want to have a quick lunch. Let me know the position of the nearest fast food restaurant.' The NLP system will understand the instruction from the user and guide the user to the recommended restaurant.

Question answering (QA) is a research sub-field in NLP which attempts to deal with a wide range of question types including: fact, list, definition, How, Why, hypothetical, semantically constrained, and cross-lingual questions. The first step is to make the machine to understand semantic expressions and extract keywords in a question.

Semantic Learning in AI

Semantic learning in AI focuses on learning the relation among terms, such as words and phrases. The word or phrase 'semantics' itself denotes a range of ideas, from the popular to the highly technical.

Feature Selection

Semantic algorithms usually first start with selecting some feature keywords. There are three basic feature selection criteria: 1) pruning of infrequent features, 2) pruning of high frequency features and 3) choosing features which have high mutual information with the target concept.

First, we prune the infrequent words from our training sets of data. This removes most spelling errors and speeds up the following stages of features selection. Second, we prune the most frequent words. This technique is supposed to eliminate noncontent words like 'the,' 'and,' or 'for.' Finally, we use the remaining words to be ranked by TFIDF technique according to their mutual information with the target concept/category.

Measuring the Significance Level among Words

After we extract the useful keywords, we try to measure the significance level among words. The Term-Frequency-Inverse-Document-Frequency (TFIDF) is an information retrieval technique commonly used as a significance indicator. This technique formulates the significance of a term/word according to its frequency in a document or a collection of documents. This work uses TFIDF to determine the weights that are assigned to individual terms. If a term t occurs in document d,

$$w_{di} = t f_{di} \times \log(N/idf_{di}) \tag{11.1}$$

where ti is a word (or a term) in the document collection, w_{di} is the weight of t_i, tf_{di} is the term frequency (term count of each word in a document) of t_i, N is the total number of documents in the collection and idf_{di} is the number of document in which t_i appears. The TFIDF technique learns a class model by combining document (vectors) into a vector space model.

Semantic Graph

We can now establish the semantic linkage among words. WordNet is an open electronic dictionary that can help form the semantic graph structure of every word/term. WordNet has already defined the weight value of different word relationship. Figure 11.4 shows a graph for the word taste. The words 'tart' and 'unpleasant' are at the same hierarchical level and have relations with each other so they are very similar.

Their distance in the tree structure can measure the similarity of two words. The ontology-based term frequency can be obtained by comparing the meanings of two terms,

$$otf_1 = tf_1 \times (1 + (1/D(t_1, t_2)))^{tf_2} \tag{11.2}$$

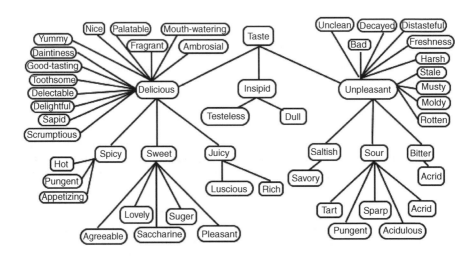

Figure 11.4   The word graph example of taste.

where t_1, t_2, are different terms; otf_1 is ontology-based term frequency of t_1; tf_1, tf_2 are the term frequency respectively to t_1 and t_2; $D(t_1,t_2)$ is the depth between t_1 and t_2. $D(t_1,t_2)$ can be calculated.

For example, assume the term frequencies of 'tart' and 'unpleasant' are 3 and 2 respectively and depth between 'tart' and 'unpleasant' is 3. The ontology-based term frequency of 'tart' will be $otf_1 = 3x(1 + (1/3))2 = 5.33$. The ontology-based term frequency of 'unpleasant' will be $otf_2 = 2x(1 + (1/3))3 = 4.67$. After adjustment of each term frequency, their term frequency value may increase. The two terms could thus become more significant after computing TFIDF.

After we get new TFIDF values, a query or an article can be simplified into a vector which contains multiple dimensions of terms and TFIDF values. Then we can measure the Euclidean distance among vectors and use clustering methods (such as K-nearest neighbor) to gather the useful data and return the result of the query.

11.3 Example of the Tourist Guide Application

In this example, we're going to implement a Tourist Guide application in iPhone which has features of positioning, providing reviews and suggestions for restaurants in the area and guiding the user to the destination. The system links to a database which contains restaurant information in Hong Kong and makes use of fuzzy logic to return the search result of favorable dining places according to the user's preference.

11.3.1 System Overview of the Tourist Guide Application

In this section, we introduce the system overview of the Tourist Guide application. Figure 11.5 describes the system overview. There are three main layers in our application.

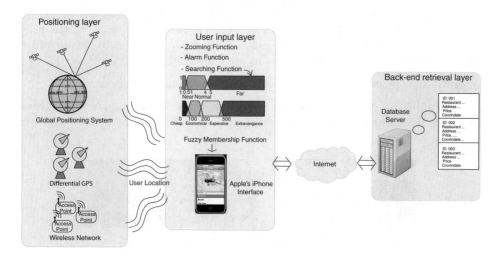

Figure 11.5   System overview of Tourist Guide application using Apple's iPhone.

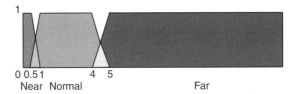

Figure 11.6 Fuzzy membership graph for distance. (A full color version of this figure appears in the color plate section.)

The first layer is the positioning layer. We use hybrid positioning technology to locate the user. iPhone 4GS has already included the features of GPS. We further extend the positioning functions indoors by Wi-Fi positioning techniques. In most cases, when the user stays in an outdoor environment, GPS would be used to estimate the user's location. When the iPhone cannot receive satellite signal or received satellite signal is very weak, the proposed system will automatically switch to Wi-Fi positioning.

The second layer is the user input layer which reads the users' query, displays the restaurant information and identifies the location in Google Maps. We implement our system using iPhone as the front-end user interface. iPhone is a touch device that can easily control the zooming function of the map. We also implement alarm functions to remind and provide suitable choices of restaurants according to users' preference.

The third layer is the back-end retrieval layer. This layer includes a location-aware database server which stores the name, address, price, type and global coordinate of restaurants.

We describe the implementation in several parts. They are the use of fuzzy logic, database structure, search engine on server side, and the iPhone app on the client side.

11.3.2   Applying Fuzzy Logic in the Tourist Guide Application

In this section, we make use of fuzzy logic to represent the distance and price. Fuzzy membership functions are used to represent the distance between the restaurant and the user location. The membership function of term set is $\mu(Distance)$={Near, Normal, Far}. Figure 11.6 shows the fuzzy membership graph in which X-axis represents the distance in kilometers and Y-axis represents the fuzzy membership from 0 to 1.

Similarly, fuzzy membership functions could be used to represent the price:

$\mu(Price)$ = {Cheap, Normal, Expensive, Extravagant} is the membership function of term set. Figure 11.7 shows the fuzzy membership graph in which X-axis represents the price in

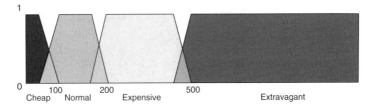

Figure 11.7   Fuzzy membership graph for price. (A full color version of this figure appears in the color plate section.)

Hong Kong dollars and Y-axis represents the fuzzy membership from 0 to 1. The trapezoid function is used to represent the fuzzy membership functions of the distance and the price.

In the Tourist Guide application, there are four conditions for selection in searching desired restaurant: Restaurant Type, Location, Price and Food Rating. Restaurant Type and Location are discrete choices, while Price and Food Rating are presented as semantic terms by utilizing fuzzy logic. Users can choose the desired option among any condition, and the Tourist Guide appication will perform an exact match on the discrete choice and fuzzy search on the semantic term. The searching result will display the most desirable restaurant at the top of the list based on the aggregated fuzzy membership score among the semantic terms.

11.3.3 Building the Database Structure

In iOS, you can use its embedded relational database, called SQLite3 to build a simple database. SQLite is a software library that implements a self-contained, serverless, zero-configuration, transactional SQL database engine. SQLite is the most widely deployed SQL database engine in the world. The source code for SQLite is in the public domain. There are many software tools that can help you to create a SQLite database. One of the popular freeware programs for creating a SQLite database is SQLite Database Browser.

SQLite Database Browser is a freeware, public domain, open source visual tool used to create, design and edit database files compatible with SQLite. It is meant to be used by users and developers who want to create databases, edit and search data using a familiar spreadsheet-like interface, without the need to learn complicated SQL commands.

You can download it from: http://sourceforge.net/projects/sqlitebrowser/.

The other efficient way is to use the http connections to retrieve the query result from the database system. In that case, we are not limited to using only SQLite but we can also use any other database systems such as Oracle, MySQL and Sybase.

Using MySQL as the Database System

In this system, we use the MySQL as a database system to store the data. The reason we pick MySQL is because it runs on more than 20 platforms including Mac OS, Linux, Windows, Solaris, HP-UX, IBM AIX, etc. The MySQL database has become the world's most popular open source database and it supports a new generation of applications built on the LAMP stack (Linux, Apache, MySQL, PHP / Perl / Python.) Many of the world's largest and fastest-growing organizations including Facebook, Google, Adobe, Alcatel Lucent and Zappos rely on MySQL to save time and money powering their high-volume websites, business-critical systems and packaged software.

There are five tables to store the restaurant information such as the location, cuisine type, price, and the restaurant rating. The attributes in each table are as follows:

Restaurant_Location
Id (int) unique id for the location
Location_name(string) the name of the location

Cuisine
Id (int) unique id for the cuisine
name(string) the name of the cuisine

Price

Id (int) unique id for the price

name(string) the name of the price range

Cuisine_Type

Shop_Id (int) unique id for the restaurant

Category_id(int) the name of the cuisine

Restaurant

shopid (int) unique id for the restaurant

name(string) the name of the restaurant

lat(double) Latitude of restaurant

lon(double) Longitude of restaurant

district_id(int) district id of restaurant

price_id price id link to price table smile(int) good rating

cry(int) bad rating

addr(string) address of restaurant

11.3.4   Setting Up the Server Side in PHP

In the server side, we use PHP language to develop the server-side system to handle the HTTP request from clients. Here is the PHP code for connecting MySQL server:

```
<?php
    $db=mysql_connect("hostname","user","password");
    mysql_select_db("database");
    mysql_unbuffered_query("SET NAMES 'utf8'");
?>
```

The above code is to connect MySQL server by *mysql_connect(hostname, user, password)* function call, and then use *mysql_select_db(database)* to select the target database. Lastly, we set the data encoding in UTF8.

When a client makes a HTTP GET request, we can get the searching parameters by using the following PHP code:

```
$district = $_GET['district'];
$price = $_GET['price'];
$page = $_GET['page'];
$type = $_GET['type'];
$rating = $_GET['rating'];
```

The above code can get the value of each parameters. There are several parameters in searching restaurant and we can use $_GET[name] to get the value of each parameter. Then we can do validation on the parameter by the following code:

```
if (isset($rating)) {
    if ($rating != "Any") {
        switch ($rating) {
        case "Excellent":
            $condition .= " and b.smile > 100";
            break;
```

```
        case "Good":
            $condition .= " and b.smile >= 50 and b.smile <100";
            break;
        case "Normal":
            $condition .= " and b.smile >= 20 and b.smile <50";
            break;
        case "Poor":
            $condition .= " and b.smile < 20";
            break;
        }
    }
}
```

After finished the checking on parameters, we can formulate and run the SQL and get the result. Here is an example of SQL script:

```
$query = "select b.name as name, b.lat as lat, b.lon as lon, a.
        location_name as district, b.addr as addr, b.smile as
        rating , b.shopid as shopid, e.name as price, d.name as
        cate from location as a, restaurant as b, categories as c,
        category_type as d, price as e where a.location_name=b.
        district and c.category_id=d.id and c.shop_id=b.shopid and
        b.price_id=e.id".$condition;
$result = mysql_query($query);
```

We use *sql_query(query)* to execute the SQL and get the result to variable *result*. The next step is to display the result:

```
while ($row = mysql_fetch_array($result))
{
echo $row['shopid']."^".$row['name']."^".$row['lat']."^".$row['lon'
        ]."^".$row['price']."^".$row['cate']."^".$row['district'].
        "^".$row['addr']."^".$row['rating']."\r\n";
}
```

We use *while ($row = mysql_fetch_array($result))* looping to fetch the record one by one from result set, and then we use echo function to output the data of each record to standard output, which is the output stream to client side.

11.3.5   Setting Up the Client-side in iPhone

The Tourist Guide application has several functions. It includes the searching function on restaurant, display restaurant annotation on the map with detail information, and alarm notification of reserved restaurant. First, we introduce the basic class for storing the restaurant information.

```
@interface restaurantInfo : NSObject <MKAnnotation>
{
    CLLocationCoordinate2D coordinate;
    NSString *title;
    NSString *subtitle;
    NSString *price;
```

```
    NSString *address;
    NSString *shopType;
    NSNumber *currentPoint;
    NSString *rating;
    itemsType itemType;
}
```

The class *restaurantInfo* is an implementation class of protocol *MKAnnotation* for storing restaurant information. It has the data member of title, subtitle, price, address, cuisine type, and rating.

```
@interface setting : NSObject <MKAnnotation>{
    CLLocationCoordinate2D coordinate;
    NSString *cuisine;
    NSString *price;
    NSString *district;
    NSDate *timealarm;
    NSString *rating;
}
```

Secondly, there is a class *setting* to save the searching parameters, which includes cuisine type, price, district, rating of the restaurant, and the alarm time.

Map View and List View Display

In the Tourist Guide application, there are two major views on the screen: Map view and List view. The map view in the upper part of the screen is used to show the location of restaurants on the map according to GPS coordinates, and we can drag and zoom the map to different regions to see the distribution of restaurants. Each restaurant is represented as an annotated object and displayed as a green pin on the map; we can tap on the green pin and the name and district of this restaurant will pop up in a small dialog box shown above the green pin.

The list view in the lower part of the screen shows the name of restaurants in this region in list format. Users can scroll the restaurant list and select the desired restaurant, and map view will move the view to focus the selected restaurant accordingly. Figure 11.8 shows the layout of the application.

Searching Function for Restaurant

We can perform searching on restaurant in the Tourist Guide application by providing different parameters, and the application will make connections to a database server and submit query request by formulating the parameters in the URL string as the code below:

```
-(void)searchRestaurant
{
    NSArray *existingpoints = mapView.annotations;
    if ([existingpoints count] > 0)
        [mapView removeAnnotations:existingpoints];

    result = [[NSMutableArray alloc] init];
    NSString *urlString = [NSString stringWithFormat:@"<PHP Server
```

Figure 11.8 Layout of Tourist Guide. (A full color version of this figure appears in the color plate section.)

```
         Link>district=%@&price=%@&rating=%@&type=%@",
         settingSearch.district, settingSearch.price,
         settingSearch.rating, settingSearch.cuisine];
urlString = [urlString stringByReplacingOccurrencesOfString:
         @" " withString:@"%20"];
NSString *locationString = [NSString stringWithContentsOfURL:[
         NSURL URLWithString:urlString] encoding:
         NSUTF8StringEncoding error:nil];
...
}
```

After the result returned from the database server, we save the result into the string variable and process it line by line, and then store each record as a *restaurantInfo* object. Each fields in the record is separated by a special character and we can use the string function *componentsSeparatedByString* to separate them into an array, so we can extract each field one by one from this array to the *restaurantInfo* object. Finally, we put the *restaurantInfo* objects into the map view annotation list. The following code demonstrate this implementation:

```
-(void)searchRestaurant
{
    ...
    NSString *locationString = [NSString stringWithContentsOfURL:[
```

```
            NSURL URLWithString:urlString] encoding:
            NSUTF8StringEncoding error:nil];
    NSArray *listItems = [locationString componentsSeparatedBy
            String:@"\r\n"];
    int total = [listItems count];
    NSLog(@"%@",urlString);
    NSLog(@"%@",locationString);
    for (int i=0; i<(total-1); i++) {
            NSString *listItemString = [listItems objectAtIndex:i];
            NSArray *listItem = [listItemString
                    componentsSeparatedByString:@"^"];

            shopCoordinate.latitude = [[listItem objectAtIndex:2]
                    doubleValue];
            shopCoordinate.longitude = [[listItem objectAtIndex:3]
                    doubleValue];

            restaurantInfo *shop = [[restaurantInfo alloc]
                    initWithCoordinate:shopCoordinate];

            [shop setTitle:[listItem objectAtIndex:1]];
            [shop setSubtitle:[listItem objectAtIndex:6]];
            [shop setPrice:[listItem objectAtIndex:4]];
            [shop setShopType:[listItem objectAtIndex:5]];
            [shop setAddress:[listItem objectAtIndex:7]];
            [shop setRating:[listItem objectAtIndex:8]];
            [shop setCurrentPoint:[NSNumber numberWithInt:i]];
            [shop setItemType:itemsTypeRestaurant];
            [result addObject:shop];
            [mapView addAnnotation:shop];
    }
    [self.tableview reloadData];
}
```

Selection in Restaurant Searching

We have a selection view to allow users to input each of the searching parameters for restaurant as shown in Figure 11.9. Each parameters has a predefined list of values for selection. When users tap the parameter, a picker view with predefined value will show up to allow users to make a selection. Here is the code listing for this implementation:

```
- (NSString *)tableView:(UITableView *)tableView
        titleForHeaderInSection:(NSInteger)section {
    switch (section) {
        case 0: {
            return @"Restaurant Type";
        } break;
        case 1: {
            return @"Price";
```

Figure 11.9   Selection in searching function. (A full color version of this figure appears in the color plate section.)

```
    } break;
    case 2: {
        return @"Rating";
    } break;
    case 3: {
        return @"District";
    } break;
    case 4: {
        return @"Set Alarm";
    } break;
    }
}
```

Alarm Notification and Providing Recommendations

In addition, an alarm function is added to remind and recommend three restaurants to the user. The code below is the implementation of alarm notification:

```
-(void) viewDidAppear:(BOOL)animated
{
    if (settingSearch.timealarm != nil) {
```

```
    NSTimeInterval tmptimeleft = [settingSearch.timealarm
            timeIntervalSinceNow];
    if (tmptimeleft > 0) {
        [setTimer invalidate];
        setTimer = [NSTimer scheduledTimerWithTimeInterval:
                tmptimeleft target:self selector:@selector(
                showAlert) userInfo:nil repeats:YES];
        NSLog(@"viewdidappear load, %d",tmptimeleft);
        NSLog(@"alarm load, %d",[settingSearch timealarm]);
    }
    }
}
```

When the time is up, it will call the showAlert method which pops up a notification dialog with three suggested restaurants. When the user clicks one of the recommendations, it shows the detailed information of the selected choice. The alert message with recommendations is shown in Figure 11.10 and the detailed information of the selected restaurant is shown in Figure 11.11, and here is the code listing for the implementation of the alert view:

Figure 11.10   Recommendation to the user.

Figure 11.11   Detailed information of the selected restaurant.

```
- (void)showAlert
{
    UIAlertView *view = [[UIAlertView alloc]
                        initWithTitle:@"Notification"
                        message:@"Recommend Restaurant"
                        delegate:self
                        cancelButtonTitle:@"Cancel"
                        otherButtonTitles:[NSString
                            stringWithFormat:@"%@",[[result
                            objectAtIndex:0] title]],
                        [NSString stringWithFormat:@"%@",[[result
                            objectAtIndex:1] title]],
                        [NSString stringWithFormat:@"%@",[[result
                            objectAtIndex:2] title]],nil];

    [view show];
    NSLog(@"ShowAlert!!!");
}
```

Showing the Guiding Trajectory

For a Tourist Guide system, apart from the searching function on restaurant, it is more useful to show the guiding trajectory and distance from the current user location to the target restaurant as the user may not be familiar with the current location. Before showing the guiding trajectory, we have to show the current user location in the map first. Developers can show the current user location in the MapView easily by enabling the *showsUserLocation* flag in MapView, and the MapView will handle all the positioning stuff by using GPS automatically. Developers can change the viewing region of the map to the current user location by using *setRegion* method. The following code shows the implementation of showing current user location and setting the viewing region in MapView:

```
mapView.showsUserLocation=TRUE;
...
MKCoordinateRegion region;
region.center = mapView.userLocation.coordinate;
[mapView setRegion:region animated:TRUE];
[mapView regionThatFits:region];
```

Although MapView provides many powerful functions such as annotation and current user location, the point-to-point trajectory and distance on map are not provided. Google Maps is a web mapping service application that offers street map, route planner for numerous countries around the world via Google Maps API, thus Google Maps is a good alternative allowing developers to show the guiding trajectory to users seamlessly. Google Maps API provides various functionalities by using different parameters, Table 11.1 is the list of supported parameters in Google Maps:

Developers can make use of a map link in Apple URL scheme to launch Google Maps in an iPhone application, by providing the corresponding parameters to display the route from current user location to target restaurant. In order to display the guiding trajectory and distance of current user location and target restaurant location, we can pass the GPS coordinate of current user location and target restaurant location to the *saddr* and *daddr* parameters in Google Maps API respectively. We can also pass the name of the restaurant in *daddr* parameter by putting the text in brackets, so that the name of the restaurant will

Table 11.1   Supported Google Maps parameters.

Parameter	Notes
q=	The query parameter. This parameter is treated as if it had been typed into the query box by the user on the maps.google.com page. q=* is not supported
near=	The location part of the query.
ll=	The latitude and longitude points (in decimal format, comma separated, and in that order) for the map center point.
sll=	The latitude and longitude points from which a business search should be performed.
spn=	The approximate latitude and longitude span.
sspn=	A custom latitude and longitude span format used by Google.
t=	The type of map to display.
z=	The zoom level.
saddr=	The source address, which is used when generating driving directions
daddr=	The destination address, which is used when generating driving directions.
latlng=	A custom ID format that Google uses for identifying businesses.
cid=	A custom ID format that Google uses for identifying businesses.

Figure 11.12 Using Google Maps to show the guiding trajectory to target restaurant. (A full color version of this figure appears in the color plate section.)

display on the target location tag in Google Maps if supported. The name string of the restaurant passed in parameter should be URL encoded with UTF8 as it is part of the URL string, otherwise the URL passing to Google Maps will be invalid and Google Map will not pop up. In the popped up window of Google Maps using the URL scheme, it will show the distance away from the current user location to the target restaurant location as well as the traveling time by using taxi, public transportation and walking. Here is the code implementation and Figure 11.12 shows the screenshots of using Google Maps to show the guiding trajectory:

```
CLLocationCoordinate2D userCoord = mapView.userLocation
        .location.coordinate;
NSString *restName = [restDetail.title
        stringByAddingPercentEscapesUsingEncoding:
        NSUTF8StringEncoding];
NSString *mapUrl = [[NSString alloc] initWithFormat:@"http://
        maps.google.com/maps?saddr=%.15f,%.15f(You)&daddr=%.15f
        ,%.15f(%@)", userCoord.latitude, userCoord.longitude,
        restDetail.coordinate.latitude, restDetail.coordinate.
        longitude, restName];
NSLog(@"Map URL = %@", mapUrl);
[[UIApplication sharedApplication] openURL: [NSURL
        URLWithString : mapUrl]];
```

Chapter Summary

In this chapter, a location-aware application with AI techniques was implemented using iPhone. You have learnt how to

▶ implement a location-aware application – Tourist Guide,

▶ apply AI techniques like fuzzy logic to search for restaurant information,

▶ build a server-end database system for processing tourist information by using MySQL and PHP,

▶ implement a map view with geotagged information that queried from server-end database,

▶ provide alarm notification and recommendation in Tourist Guide application,

▶ show the guiding trajectory from point to point by using Google Maps.

Location-aware applications are very practical for our daily lives, such as searching restaurants with detailed information and guiding trajectory information using location-aware information covered in this chapter; further enhancement can enrich the functionalities to provide more interaction with people and rich content with video. The next chapter will discuss these enhancements and provide detailed implementation to achieve them.

Chapter 12

Beyond Positioning: Video Streaming and Conferencing

Remembering that you are going to die is the best way I know to avoid the trap of thinking you have something to lose. You are already naked. There is no reason not to follow your heart.

Steve Jobs
2005

Chapter Contents

- ► What is video streaming and how does it work?

- ► What is location-aware video streaming?

- ► What is video conferencing and how does it work?

- ► What are the open problems of video streaming and conferencing in mobile environment?

- ► Implementation of video streaming and conferencing application in iPhone

Video streaming and conferencing are not new concepts to many people. The emergence of high-speed wireless network technologies such as general packet radio service (GPRS), CDMA2000 1X, third-generation (3G) mobile networks offers new opportunities for expanding the cell phone capabilities, not just a voice communication device but the effective delivery of rich multimedia content and face-to-face video conferencing. A significant amount of Internet traffic is generated by video streaming services, such as YouTube, FindVideos, and vSocial, or some sites provide video programmes like LuluTV, TVB, PPTV, Hong Kong Apple Daily News, BBC iPlayer, TV2 and NRK.

Introduction to Wireless Localization: With iPhone SDK Examples, First Edition. Eddie C.L. Chan and George Baciu.
© 2012 John Wiley & Sons Singapore Pte. Ltd. Published 2012 by John Wiley & Sons Singapore Pte. Ltd.

With the push technology in Web 3.0, location-based information will be pushed to users' smartphones when they reach the trigger location. Hangout video conferencing requests can be sent to friends nearby. Restaurant information with a video showing the dining place will be pushed to the users' smartphones when they come across the restaurant. Location-based videos related to weather and traffic conditions help the users to select a better route to their meeting places.

In this chapter we will take a look at the technology background of video streaming and conferencing. Then we extend the previous tourist guide iPhone application to support video streaming and conferencing.

12.1   What is Video Streaming?

Video-streaming is a real-time transport of a live or stored video. A user does not need to download the entire video file, but just click to view the video in a real-time and true on-demand manner. In order to do this, the bandwidth should be fast enough to download parts of the video, such that they are being decoded and played out simultaneously without lagging. Due to its real-time nature, video parts should be downloaded and decoded sequentially as long as a user plays the video. Generally, there are three types of video streaming: point-to-point, multicast, and broadcast.

12.1.1   Point-to-point Video Streaming

Point-to-point video streaming is the simplest application which one video streaming server directly connects to one single client. Examples of daily life point-to-point or one-to-one applications are video phone call and unicast video streaming. During this point-to-point streaming, the server will transfer the streamed video to the client and the client may provide feedback to the server such that the server can adjust the content or speed of video streaming according to feedback. The client can disable this feedback process, so the server has limited knowledge of the client. Figure 12.1 shows point-to-point video streaming.

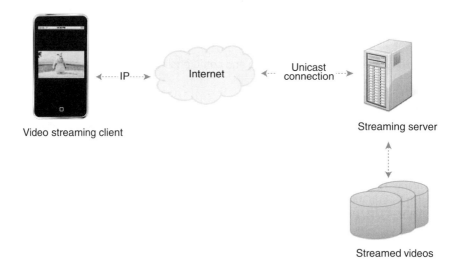

Figure 12.1   Point-to-point video streaming.

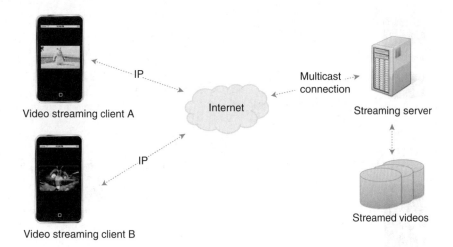

Figure 12.2   Multicast video streaming.

12.1.2   Multicast Video Streaming

Multicast video streaming is a one-to-many application by which normally one server connects to three or four clients. Instead of establishing a single unicast connection to each client, the server side establishes a multicast connection to many clients which sends a single stream from the source to a group of recipients. Examples of multicast video streaming are IP-Multicast over the Internet and application-layer multicast via overlay networks. Multicast protocols were developed to reduce the data replication that occurs when many recipients receive unicast content streams independently. One potential disadvantage of multicasting is the loss of video on demand functionality. Continuous streaming usually allow clients to control playback. However, this problem can be mitigated by elements such as caching servers and buffered media players. Figure 12.2 shows multicast video streaming.

12.1.3   Broadcast Video Streaming

Broadcast video streaming is a one-to-all application by which one server delivers all popular video contents to all clients. One of the major characteristics of broadcast video streaming is that the system must be designed to provide every intended client with the required video content. Since different clients may be interested in different video content, the system is often designed for the lowest traffic demand. To increase the performance of the broadcast video streaming, peer-to-peer sharing techniques are usually adopted such that clients can share the video content among themselves and the traffic load of the server can be significantly reduced. Typical peer-to-peer broadcast examples are playing video contents in PPTV and PPStream applications. Figure 12.3 shows broadcast video streaming.

12.2   Networks and Formats in Video Streaming

The implementation of mobile streaming services needs to cope with different access networks. Mobile roaming allows the user to stay connected with multiple networks. The technologies of mobile communication system have evolved from 1G, 2G, 3G to 4G and from Wireless LAN to Broadband Wireless Access to 4G. 4G may be considered to include

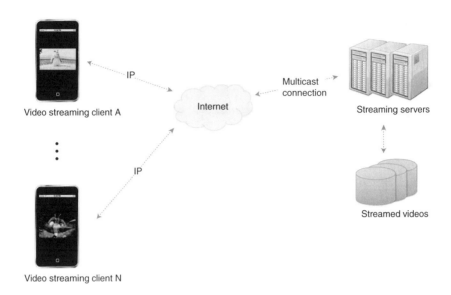

Figure 12.3   Broadcast video streaming.

3GPP LTE/LTE-Advanced, 3GPP2 UMB, and Wi-MAX based on IEEE 802.16 m. Figure 12.4 shows the data rate varying across different networks.

Streamed video content should be efficiently delivered by choosing the best connections. For example, when a user watches video sport updates in the coffee shop, he/she connects to two networks, which are the 300 Mbps Wi-Fi network provided by the coffee shop and the

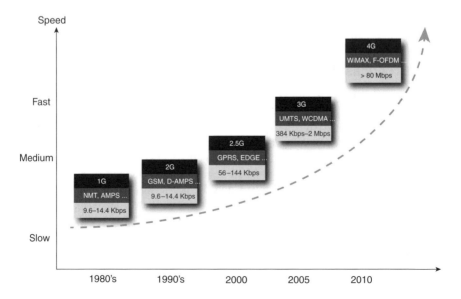

Figure 12.4   Evolution of radio access technologies.

general packet radio service (GPRS) provided by the internet service provider. The mobile streaming application should choose to use the Wi-Fi network and because he/she could get the content delivered at a higher bit rate than using a GPRS network.

The Third Generation Partnership Project (3GPP) is currently addressing mobile streaming standardization which includes both protocol and codec. The 3GPP mobile streaming standard is the most mature standardization activity in the telecommunication field, and all major mobile telecommunication equipment providers support it.

The 3GPP standard specifies protocols in video streaming:

- ▶ Real-time Streaming Protocol (RTSP) , Real-time Transport Protocol (RTP) and the Real-time Transport Control Protocol (RTCP) are specifically designed to stream media over networks

- ▶ Microsoft Media Server (MMS) is the name of Microsoft's proprietary network streaming protocol used to transfer unicast data in Windows Media Services

- ▶ Apple's HTTP Live Streaming Protocol (HLS)

- ▶ Synchronized Multimedia Integration Language (SMIL 3.0) for interactive multimedia presentations, and

- ▶ HTTP and TCP for transporting static media such as session layouts, images, text

The video and audio streaming codecs are:

- ▶ H.264 video, also called MPEG-4 Advanced Video Coding or H.264/AVC,

- ▶ VP8, an open video compression format released by Google,

- ▶ FLV, WebM, ASF, ISMA encoded in a container bitstream, and

- ▶ MP3, Vorbis, AAC audio format.

Figure 12.5 shows the video streaming platform using different protocols and networks. The video streaming clients roam with different network such as Wi-Fi, 3G or GPRS and send the HTTP request to have a SMIL presentation from a web server. Within the SMIl presentation, video streaming clients find links to the streaming content in the streaming servers.

12.3 How Does Video Streaming Work?

A video streaming platform provides video content in streaming servers where clients request for viewing video contents in an ad hoc and real-time manner via a variety of access networks, such as HTTP, Wi-Fi, Wi-MAX 3G, 4G etc.

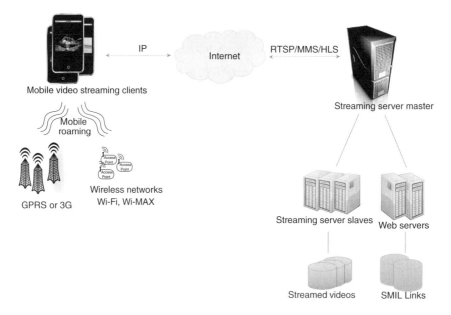

Figure 12.5   Video streaming platform.

12.3.1   Traditional Video Streaming

In the traditional video streaming, clients finds links to the streaming content in a wired network. The streaming content is in Synchronized Multi-media Integration Language (SMIL) presentation which is a W3C recommended XML markup language for describing multimedia presentations. It defines markup for timing, layout, animations, visual transitions, and media embedding, among other things. With this information, video contents will be fetched and streamed from web server to clients via HTTP connections. The video content is compressed before the transmission such that the size of content is smaller and can be downloaded faster. Clients receive the compressed video content and decode the video sequentially. The video contents are being played in the front-end process and parts of video content are downloaded sequentially inside the buffer at the background process, so that clients play the streamed video without lagging.

Nowadays, users can view streamed videos likes news, sports update or even TV programmes and movies on mobile devices while taking trains, buses, boats and etc. Mobile video streaming very often operates in a fluctuating bandwidth network and must support a variety of access networks and terminals and provide the services in an uniform way. However, the traditional video streaming does not support automatic quality and bit-rate adaptation and is not reliable to deal with the significant variations in bandwidth.

12.3.2   Adaptive Video Streaming

Adaptive video streaming is based on the traditional streaming to cope with varying resource availability, temporary connection loss, appearance of new networks, high error rates, insufficient channel capacity and etc. It allows automatic change in the video formats and in the bit-rate of transferring adaptively following the bandwidth fluctuations in the network.

Using adaptive video formats and bit-rates often means to sacrifice the video quality and maintain to play the video content without lagging. The adaptive video codecs (formats), like Multiple Description Coding (MDC), Scalable Video Coding (SVC) and scalable MPEG (SPEG) can be changed dynamically to follow the oscillating bandwidth, giving a large advantage over traditional streams that are frequently interrupted due to buffer under-runs or data loss.

The iPhone receives the streamed video via Apple's HTTP Live Streaming Protocol with Quicktime and the Darwin Streaming server at the back. Apple's HTTP Live Streaming is an HTTP-based media streaming communications protocol as part of their QuickTime X and iPhone software systems. It works by breaking the overall stream into a sequence of small HTTP-based file downloads, each download loading one short chunk of an overall potentially unbounded transport stream. As the stream is played, the client may select from a number of different alternate streams containing the same material encoded at a variety of data rates, allowing the streaming session to adapt to the available data rate. At the start of the streaming session, it downloads an extended M3U playlist containing the metadata for the various sub-streams which are available.

Windows-based phones can use Microsoft's Smooth Streaming to enable adaptive streaming of media to Silverlight and other clients over HTTP. Smooth Streaming provides a high-quality viewing experience that scales massively on content distribution networks, monitors the download speed of video segments and dynamically adapts to resource availability changes by switching between video segments coded in different qualities and bit-rates.

However, it is very common that the network connection goes down for a long period of time, for example, users driving the car across a tunnel or far away from the nearest mobile station. It would be better if we can know the network availability in advance along its intended route, so that the system can decide to when and how much data to retrieve, and which bit-rates to use.

12.4   Location-aware Video Streaming

Location-aware video streaming retrieves the network and bandwidth availability information from location-based bandwidth database. With this information, the streaming application can decide to when and how much data to retrieve, and which bit-rates to use. The location-based bandwidth database stores the network and bandwidth information across different locations on previous usages.

The client can request network and bandwidth availability information along the predictable sequence of locations. For example, the route can be known a priori when the client takes buses or trains which enables fairly good predictions of the times when locations are visited. The video streaming client can search for the lookup service that returns the positions of available networks and a sequence of bandwidth samples for each point listed in the path description using the GPS-based lookup service. These data are then used to predict and prepare for connectivity gaps, and to smooth out the video quality along a route with highly fluctuating bandwidth. Then the video streaming client can calculate the estimated number

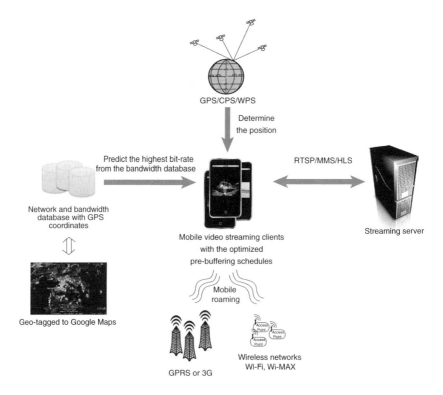

Figure 12.6   Location-aware video streaming architecture.

of bytes that it can download during the remaining time of the trip. Figure 12.6 shows the location-aware video streaming architecture.

In order to apply the location-aware streaming, a detail surveying of network bandwidth availability is performed and then the information collected in the bandwidth survey is used to derive algorithms to monitor the video streaming bit-rate and quality. In the following paragraphs, we will describe these two major steps.

12.4.1   Building the Location-Based Bandwidth Lookup Database

To build a comprehensive location-based bandwidth lookup database, intensive manual surveys of network availability and their bandwidths are performed. During the survey, the participants (users) need to go through every possible route to measure the performance of video streaming. For example, if the location-based lookup database is built within two district areas, all possible public transportation routes (such as metro railway, bus, tram etc.) that connected two districts should be all measured.

The participants need to perform bulk data transfers (i.e. watching a streamed video) throughout the routes. As mobile roaming technology allows participants to stay connected more than one networks (such as Wi-Fi, GPRS and 3G together), the video streaming application should be able to automatically switch and test to the fastest connected network.

The participants also need to locate themselves precisely using GPS, CPS or WPS throughout the routes, such that the collected data can be marked and recorded with the standardized geographic information system (GIS) information, for example, tagging the collected data with the precise GPS coordinates in Google Maps. The collected data includes network type and its id, time, GPS coordinates and observed streaming performance metrics like bandwidth, round-trip time and packet loss rate.

Each possible route should be taken multiple measurements at different time intervals, for example, morning and evening period. Because the network availability may be fluctuated according to the number of people using the network at different time intervals. Finally these throughput data is recorded and stored to the lookup database server.

12.4.2 Location-based Bit-rate and Quality Monitoring

Using the collected survey data, a predictive algorithm is derived that the streamed video are not interrupted due to buffer under-runs or data loss. The predictive algorithm makes use of the observed bandwidth, round-trip time and packet loss rate in the lookup database to calculate the highest video streaming bit-rate and adjust the quality of the video. The video streaming application then uses the predicted bit-rate value to estimate the amount of data that can be downloaded. This helps to efficiently manage the downloading schedule. For example, when the bandwidth is predicted to be fast and available (open area with good connection), we can schedule to pre-buffer more video. On the other hand, when the network is slow or unreachable (inside tunnel with poor connection), the video streaming application can stop requesting to download and reduce the usage of power. To prevent the error between the prediction and the real bandwidth availability, the adaptive video streaming technique is also usually used to calculate the highest bit-rate value which serves as the upper bound of predicted bit-rate value.

The pricing is another issue to be considered. With the network provider information in the lookup database, we can try to come up with a hybrid downloading scheme which is cheaper and also provide a stable video streaming service.

Open Problems in Location-aware Streaming

Location-aware video streaming is still under development progress because building a bandwidth lookup service across a huge area, like campus site, district or even city requires massive amount of manual site-surveying process to record the availability of network and bandwidth information. Different research groups in the world have established different policies, schema and standards for collecting location-based information and still have no censuses yet.

12.5 What is Video Conferencing?

High speed mobile connectivity has become more widely available and most of smartphones today have a front-facing camera which support video calls. Video conferencing (video call) provides a face-to-face communication between two or more people, enables people to connect with friends, stays closer to family around the world in a way that is fun, easy, spontaneous, and fits the moment.

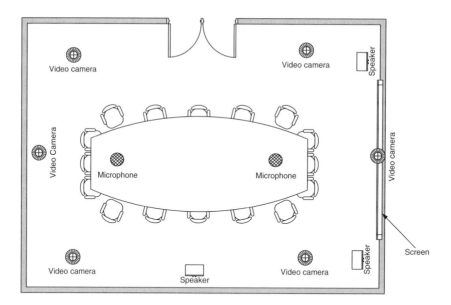

Figure 12.7   Video conference room settings.

Similar to the video streaming technology, there are three main types of video conferencing: one-to-one, multi-point and broadcast. The private user usually operates one-to one video conferencing call. In business, offices may be installed multiple video conferencing cameras to have a full view of the meeting room. Microphones and speakers are configured well to receive users' conversations. Users can switch to see from different perspectives using different cameras. Figure 12.7 shows an example of video conference room settings.

The source of video conferencing must be captured and encoded for real-time communication, whereas the source of video streaming can be either pre-recorded or real-time. The source of video conferencing has a time-bounded usefulness. For example, if there happens some delays on a video call, then the delayed video part is useless. Capturing, encoding, transferring, receiving, decoding and displaying are all in a timely manner and they process on-the-fly. The maximum acceptable latency is around 150 ms.

Other than that, from the technical point of view, the challenges faced in video conferencing are almost near the same as video streaming. It also needs to cope with the unstable resource availability, temporary connection loss, appearance of new networks, high error rates, insufficient channel capacity etc. The techniques to adjust the video bit-rate and quality can also be applied to video conferencing.

What are the Partitioners?

On June 7, 2010, FaceTime was released by Apple at the WWDC 2010 which is a one-to-one video calling software application supported only in iOS devices and Macintosh computers equipped with FaceTime Cameras. FaceTime in iPhone is an integral part of the phone call application. Users can access to their contact file in iPhone to initiate a HD FaceTime call only in Wi-Fi connection. If using a Macintosh version of FaceTime, users need to register

Apple ID email addresses to initiate a FaceTime call. Users will receive a push notification when they get a call.

FaceTime is based on numerous technologies:

- ▶ H.264 (video) and AAC (audio),

- ▶ SIP IETF signaling protocol for VoIP,

- ▶ STUN, TURN and ICE IETF technologies for traversing firewalls and NAT, and

- ▶ RTP and SRTP IETF standards for delivering real-time and encrypted media streams for VoIP.

Tango was founded in September 2009 which provides the alternative one-to-one video conferencing solution and supports platforms, including PC with webcams, iOS and Android devices. In the contrast of FaceTime, Tango supports the video conferencing experience over 3G, 4G, and Wi-Fi. In order to use the Tango service, the user need to register for a Tango account and install the Tango software. Similarly, the user can initiate a video call by tapping the name (users already installed Tango) in the Tango contact list. It supports the push notification even the user shuts down the Tango application.

Skype was founded in 2003 which also provides both one-to-one and group video conferencing solution. In addition, it also supports to both traditional landline telephones and mobile phones for a fee using a debit-based user account system. It can be installed and run almost every platform including PC, MAC, Linux with webcams, iOS, Android and Symbian devices with front-facing cameras. Similarly, the user need to register a Skype account to initiate a video call by tapping the name (users already installed Skype) in the Skype contact list. Unlike other VoIP services, Skype is a peer-to-peer system rather than a client-server system. Users need to turn the Skype application in background to receive a push notification when they get a call.

Google+ was launched on June 28, 2011 which supports one-to-many video conferencing with the features of social networking and positioning services. This application introduces a new way of video-conferencing based on social networks and locations. Google+ allows users to communicate the friends nearby via video calls. This video conferencing is called 'Mobile Hangouts' and supports Android 2.3+ devices with front-facing cameras, PC, MAC, Linux with webcams. It is an example of location-aware video-conferencing as described in Figure 12.8.

After we have a very thorough knowledge about video conferencing and streaming. Let's implement the location-aware video streaming and conferencing applications in iPhone.

12.6 Implementation of Video Streaming in iPhone

iPhone is powerful in processing multimedia data as it has built-in multimedia processor and complete AV foundation framework. For video streaming in iPhone, we can use the *MPMoviePlayerController* object in Media Player framework or *UIWebView* object for web-based media. In our Tourist Guide application, we implemented a video streaming player by using the *MPMoviePlayerController* object. *MPMoviePlayerController* manages

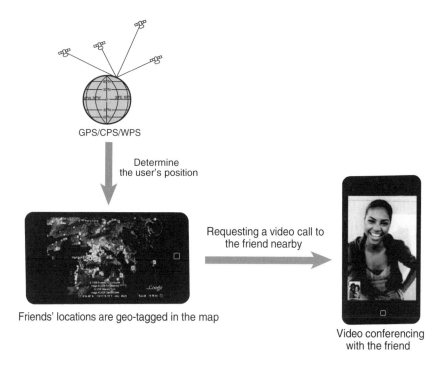

GPS/CPS/WPS

Determine
the user's position

Requesting a video call to
the friend nearby

Friends' locations are geo-tagged in the map

Video conferencing
with the friend

Figure 12.8   Location-aware video conferencing.

the playback of a movie from a file or a network stream. Playback occurs in a view owned by the movie player and takes place either fullscreen or inline. Here is the code implementation of using *CustomMoviePlayerViewController* player to play the streaming video with the detailed information of the selected restaurant in Tourist Guide application:

```
- (void)loadMoviePlayer
{
    NSURL *path = [NSURL URLWithString:@"http://127.0.0.1:1935/vod/
            mp4:sample.mp4/playlist.m3u8"];
    // Create custom movie player
    moviePlayer = [[CustomMoviePlayerViewController alloc]
            initWithPath:path];
    // Show the movie player as modal
    [self presentModalViewController:moviePlayer animated:YES];
    // Prep and play the movie
    [moviePlayer readyPlayer];
}
```

HTTP Live Streaming

From the above code implementation, we initiate the movie player with a HTTP URL, which then presents the player in modal view and starts playing. The *MPMoviePlayerController* object supports HTTP Live Streaming Protocol (also known as HLS), it is a HTTP-based media streaming communication protocol implemented by Apple Inc. as part of their

QuickTime X and iPhone software systems. It works by breaking the overall stream into a sequence of small HTTP-based file downloads, each download loading one short chunk of an overall potentially unbounded transport stream.

In the HTTP URL of streaming video, we specify the *playlist.m3u8* file for streaming. The *m3u8* file is the unicode version of *m3u*, *m3u* is a computer file format that stores multimedia playlists. An *m3u* file is a plain text file that specifies the locations of one or more media files that the media player should play.

Setup of HLS Server

In order to demonstrate a complete implementation of video streaming, we will implement the streaming client in the Tourist Guide application and setup a HLS server. Wowza Media Server is an on-demand live streaming server developed by Wowza Media Systems. It supports various kinds of streaming protocols like Real Time Streaming Protocol (RTSP), Real-time Transport Protocol (RPT), especially support for Apple HTTP Live Streaming protocol for iOS devices on H.264 media format as in our demonstration. Here is the instruction to setup the Wowza Media Server on Macintosh PC.

1. First of all, we download the Mac OS X installation of Wowza Media Server version 3.0.2 from the website of Wowza Media Systems:

http://www.wowza.com/store.html

2. Apply the free developer license by providing personal information and email, and Wowza Media Systems will send the developer license key to your email for activating the server in this link:

http://www.wowza.com/developer.html

3. After downloaded the DMG image of Wowza Media Server version 3.0.2 (WowzaMediaServer-3.0.2.dmg), double click this image file to mount it in Finder, and a new Finder window will pop up and there is a *WowzaMediaServer-3.0.2.pkg* installation package in this image as shown in Figure 12.9.

4. Double click the *WowzaMediaServer-3.0.2.pkg* installation package to start the installation, you will see the installation screens and click proceed buttons to complete the installation as in Figure 12.10.

5. After finished the installation, go to the *Application* in Finder. You will find an application named *Wowza Media Server 3.0.2* added. Click into it, there is a *Wowza Startup.app* application for you to start up the media server.

WowzaMediaServe
r-3.0.2.dmg

WowzaMediaServer-3.0.2.
pkg

Figure 12.9   DMG and PKG Files of Wowza Media Server.

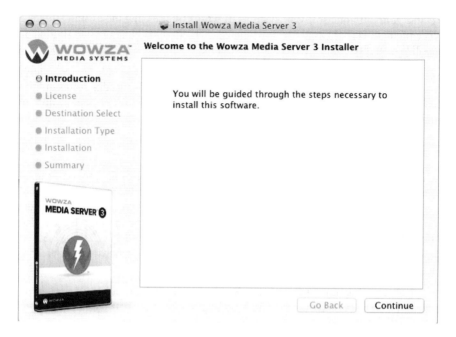

Figure 12.10   Installing Wowza Media Server.

6. Double click *Wowza Startup.app* application to startup the Wowza Media Server version 3.0.2, you will get the terminal console as in Figure 12.11. You are required to input the license key for the first time startup and input the developer license key received in email in Step 2 to continue the startup as in Figure 12.12. Finally, the Wowza Media Server is

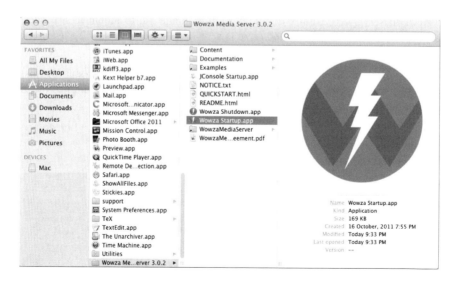

Figure 12.11   Application of Wowza Media Server.

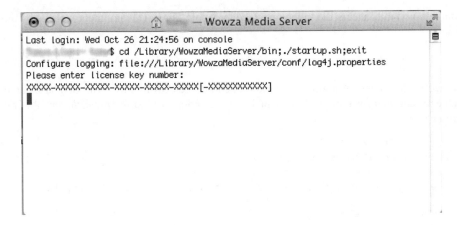

Figure 12.12   Entering the license key of Wowza Media Server.

alive as shown in Figure 12.13. It is ready for video streaming as sample.mp4 file which is already setup for your testing by default. This sample streaming video file will be accessed in our Tourist Guide application on iPhone simulator by using HLS on this URL:

http://<stream server ip>:1935/vod/mp4:sample.mp4/playlist.m3u8

This is the quick setup of Wowza Media Server for the demonstration of video streaming in Tourist Guide application, please refers to the server documentation for detailed setup on streaming.

Implementation of Streaming Movie Player

After the HLS server is setup, we can access the video by HLS protocol by *MPMoviePlayerController* in Tourist Guide application. If developer do not want to setup the HLS server

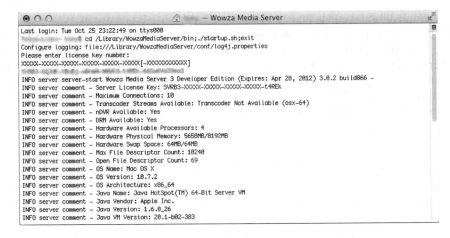

Figure 12.13   Successful Startup of Wowza Media Server.

yourself, you may access the video provided by Apple Image Development website by this link:

> *http://devimages.apple.com/iphone/samples/bipbop/bipbopall.m3u8*

In the real-life practical application, the link should be downloaded from database server in the restaurant information. Here is the implementation of *CustomMoviePlayerViewController* player:

```
#import <UIKit/UIKit.h>
#import <MediaPlayer/MediaPlayer.h>

@interface CustomMoviePlayerViewController : UIViewController
{
    MPMoviePlayerController *mp;
    NSURL *movieURL;
}
- (id)initWithPath:(NSURL *)url;
- (void)readyPlayer;
@end


@implementation CustomMoviePlayerViewController
- (id)initWithPath:(NSURL *)url
{
    // Initialize and create movie URL
    if (self = [super init])
    {
        movieURL = url;
    }
    return self;
}
...
@end
```

We will initialize the movie play and associated URL when it is created, and we pass the NSURL to indicate the address or playlist of streaming video. We create the player and setup the notification where the movie player work on different iOS versions. The movie player should call to the response to selector as Apple recommends this way to check for feature availability. For iPhone running 3.2 or above, there is a *loadstate* method, and we can set the player to fullscreen view. Here is the code implementation and Figure 12.14 shows the video streaming in Tourist Guide:

```
- (void) readyPlayer
{
    mp =   [[MPMoviePlayerController alloc] initWithContentURL:
              movieURL];

    if ([mp respondsToSelector:@selector(loadState)])
    {
        // Set movie player layout
        [mp setControlStyle:MPMovieControlStyleFullscreen];
```

(a) Video streaming link. (b) Pop up the player window in fullscreen in restaurant details.

(c) Slide the timeline of video to stream on specific time.

Figure 12.14　Video streaming in Tourist Guide.

```
    [mp setFullscreen:YES];
    // May help to reduce latency
    [mp prepareToPlay];
    // Register that the load state changed (movie is ready)
    [[NSNotificationCenter defaultCenter] addObserver:self
        selector:@selector(moviePlayerLoadStateChanged:)
        name:MPMoviePlayerLoadStateDidChangeNotification
        object:nil];
}


// Register to receive a notification when the movie has
        finished playing.
[[NSNotificationCenter defaultCenter] addObserver:self
        selector:@selector(moviePlayBackDidFinish:)
        name:MPMoviePlayerPlaybackDidFinishNotification
        object:nil];
}
```

Function *moviePlayerLoadStateChanged* is the selector method when the movie player
state changed, it first removes the observer, then change the current view to landscape and
fullscreen mode. Finally play the video.

```
- (void) moviePlayerLoadStateChanged:(NSNotification*)notification
{
    // Unless state is unknown, start playback
    if ([mp loadState] != MPMovieLoadStateUnknown)
    {
        // Remove observer
        [[NSNotificationCenter defaultCenter] removeObserver:self
            name:MPMoviePlayerLoadStateDidChangeNotification object:
                nil];

        // When tapping movie, status bar will appear, it shows up
        // in portrait mode by default. Set orientation to
        // landscape
        [[UIApplication sharedApplication] setStatusBarOrientation:
                UIInterfaceOrientationLandscapeRight animated:NO];
        // Rotate the view for landscape playback
        [[self view] setBounds:CGRectMake(0, 0, 480, 320)];
        [[self view] setCenter:CGPointMake(160, 240)];
        [[self view] setTransform:CGAffineTransformMakeRotation
                (M_PI / 2)];
        // Set frame of movieplayer
        [[mp view] setFrame:CGRectMake(0, 0, 480, 320)];

        // Add movie player as subview
        [[self view] addSubview:[mp view]];
        // Play the movie
        [mp play];
    }
}
```

Function *moviePlayBackDidFinish* is the selector method when movie player is finished,
it shows the status bar and reset the style of status bar first, and then remove the observer.
Finally quit the modal view.

```
- (void) moviePlayBackDidFinish:(NSNotification*)notification
{
    [[UIApplication sharedApplication] setStatusBarHidden:NO];
    [[UIApplication sharedApplication] setStatusBarStyle:
            UIStatusBarStyleDefault];
    // Remove observer
    [[NSNotificationCenter defaultCenter] removeObserver:self
        name:MPMoviePlayerPlaybackDidFinishNotification object:nil];
    [self dismissModalViewControllerAnimated:YES];
}
```

12.7   Implementation of Video Conferencing in iPhone

With the launch of FaceTime, video conferencing in iPhone is no longer a sophisticated task as it is built-in in every newly released iPhone and iDevice. Apple provides URL scheme API to allow applications to invoke FaceTime inside the application seamlessly. Before we go into the URL scheme for FaceTime call, we should think of how to incorporate the friend object in our application, so that we can make FaceTime call to a friend. A very interesting idea in our Tourist Guide application that a friend object is presented in the map view with geo-tagged information together with the restaurant information, so that we can figure out where my friend is, and who is near me, and then tap the friend object on the map to make FaceTime call directly. The idea of the geo-tagged friend object is that the Tourist Guide user will register with our centralized database server with his or her name, telephone number (FaceTime ID) and photo first, and the Tourist Guide application will report the current user location to the database server periodically, say every 2 minutes; all this user location information in our centralized database server will send to the Tourist Guide application periodically also, so that the application can display this information as a geo-tagged friend annotation object on the map.

Similar to the restaurant annotation object in our Tourist Guide application before, the friend annotation object is an implementation of protocol *MKAnnotation* class with information of telephone and image object. Here is the code listing of the friend annotation object:

```
@interface FriendAnnoInfo : NSObject <MKAnnotation>

@property (nonatomic) CLLocationCoordinate2D coordinate;
@property (nonatomic, copy) NSString *title;
@property (nonatomic, copy) NSString *subtitle;
@property (nonatomic) itemsType itemType;
@property (nonatomic, retain) NSNumber *currentPoint;

@property (nonatomic, retain) NSString *telephone;
@property (nonatomic, retain) UIImage *image;

@end
```

After we get the friend information from the centralized database server, we can initialize *FriendAnnoInfo* objects with the location, name, image and telephone information, and then add them to the map view. In the map view, we can display the geo-tagged friend object to the his or her current location on the map together with his or her photo by the following code implementation and Figure 12.15 shows the geo-tagged friend object displayed on the map view with photo:

```
- (MKAnnotationView *) mapView:(MKMapView *)mapView
        viewForAnnotation:(restaurantInfo *) annotation {
    ...
    if ([annotation isKindOfClass:[FriendAnnoInfo class]])
    {
        MKPinAnnotationView *annView=[[MKPinAnnotationView alloc]
                initWithAnnotation:annotation reuseIdentifier:nil];
        annView.pinColor = MKPinAnnotationColorPurple;
```

Figure 12.15   Geo-tagged friend objects on map view. (A full color version of this figure appears in the color plate section.)

```
UIButton *myDetailButton = [UIButton buttonWithType:
        UIButtonTypeCustom];
myDetailButton.frame = CGRectMake(0, 0, 23, 23);
myDetailButton.contentVerticalAlignment =
        UIControlContentVerticalAlignmentCenter;
myDetailButton.contentHorizontalAlignment =
        UIControlContentHorizontalAlignmentCenter;
// Set the image for button
[myDetailButton setImage:[UIImage imageNamed:@"button_right.
        png"] forState:UIControlStateNormal];
[myDetailButton addTarget:self action:@selector(showLinks:)
        forControlEvents:UIControlEventTouchUpInside];
// Set the button as the callout view
annView.rightCalloutAccessoryView = myDetailButton;
annView.animatesDrop=TRUE;
annView.canShowCallout = YES;
annView.calloutOffset = CGPointMake(-5, 5);

FriendAnnoInfo *friend = [friendSet objectForKey:annotation.
        currentPoint];
```

```
    UIImageView *imageView = [[UIImageView alloc] initWithImage:
        friend.image];
    myDetailButton.tag = [annotation.currentPoint intValue];
    [annView addSubview:imageView];
    return annView;
  }
  ...
}
```

After we displayed the friend annotation object on the map view, we can implement the click event handler of the friend annotation object to make FaceTime call to the target friend by means of the URL scheme. We can pop up an alert dialog of making a FaceTime call to allow the user to choose before making the FaceTime call. For the URL scheme of making the FaceTime call, the URL is started with *facetime://* and is followed with the telephone number or Apple ID for that friend. Figure 12.16 shows the screenshots of the FaceTime call.

Here is the code for making a FaceTime call:

```
- (IBAction)showLinks:(id)sender {
    int nrButtonPressed = ((UIButton *)sender).tag;
```

Figure 12.16   Making FaceTime call in Tourist Guide.

```
    ...
    FriendAnnoInfo *friend = [friendSet objectForKey:[NSNumber
            numberWithInt:nrButtonPressed]];
    NSString *alertStr = [[NSString alloc] initWithFormat:@"Make
            FaceTime call to %@?", friend.title];
    UIAlertView *redirectAlert = [[UIAlertView alloc] initWithTitle:
            @"FaceTime Friend" message:alertStr delegate:self
            cancelButtonTitle:@"Cancel" otherButtonTitles:@"OK",
            nil];
    redirectAlert.tag = nrButtonPressed;
    [redirectAlert show];
    ...
}

-(void)alertView:(UIAlertView *) alertView clickedButtonAtIndex:(
        NSInteger) buttonIndex
{
    ...
    if (buttonIndex == 1) {
        FriendAnnoInfo *friend = [friendSet objectForKey:[NSNumber
                numberWithInt:alertView.tag]];
        NSString *facetimeStr = [[NSString alloc] initWithFormat:@"
                facetime://%@", friend.telephone];
        [[UIApplication sharedApplication] openURL: [NSURL
                URLWithString: facetimeStr]];
    }
    ...
}
```

Chapter Summary

This chapter has covered fundamental theories and technical problems of video streaming and conferencing. Most interestingly, we demonstrate and implement the idea of location-aware video streaming and conferencing in iPhone.

In the implementation of video streaming, we have demonstrated a complete server and client video streaming implementation. We setup a HLS server using Wowza Media Server to provide streaming source and use *MPMoviePlayerController* in AV foundation framework to play streaming video seamlessly in the Tourist Guide application, so that enriched multimedia content can be brought to users in a Tourist Guiding system.

In the implementation of video conferencing, we use iPhone's built-in FaceTime function to make a video conference call with a friend, and the location of the geo-tagged friend objects can be displayed on the map view according to their current location.

APPENDIX A
STARTING THE iOS SDK

What is in this appendix chapter?

▶ Getting the iOS SDK

▶ What tools are in the iOS SDK?

▶ Introduction to Xcode

Before you start to experience of programming in iPhone, you should prepare several things. First, you should have an Intel-based Macintosh computer. And its operating system should be Snow Leopard (OS X 10.6.6 or above) or Lion (OS X 10.7 or above). After you have a Mac OS system, the next step is to get the iOS SDK. The iOS SDK was officially announced on March 6, 2008. Before that day, the kit had no name. Apple only said that iPhone runs on 'OS X,' which is a reference to iPhone OS's parent, Mac OS X. The first beta release version of the SDK is v.1.2. Two months later, the iOS SDK v2.0 beta 6 could be used in Mac OS X. Up to October 2011, the version of iOS SDK is 4.2. This is the tool chain used for developing on the iPhone platform. It includes over thousands of APIs for developing iPhone applications.

A.1 Getting the iOS SDK

To get the iOS SDK, you must first sign up to an Apple ID account. You can register directly under the iPhone Developer Center's website:

http://developer.apple.com/iphone

After you successfully sign up and login your account, you can see the link for downloading the latest version of iOS SDK. You should see the webpage as shown in Figure A.1. You can select to download either Snow Leopard or Lion version of SDK. You simply click the link and get the iOS SDK by Safari or download the iOS SDK via App Store.

You could download the free iOS SDK. Although you can test your program in the iPhone simulator, the free option will not allow you to deploy your applications onto your actual iPhone (or any iOS devices). If you would like to distribute your iPhone application on the App Store, you could join an iOS Developer Program. There are three types of program – Individual, Company, and Enterprise Programs. Individual Program is designed for developers who can distribute the free or commercial applications on the App Store. It costs US$99. Company Program is designed for companies that can distribute the free or commercial applications on the App Store and also have the ability to create developer team. Enterprise Program is designed for companies that have 500 or more employees to create proprietary in-house applications. It costs US$299.

No matter whether the user joins Individual, Company or Enterprise Program, the developer center will provide a very considerate code-level technical support. Also if you join any of these three programs, you can test your application in different version of iOS devices. You can enroll the developer program as shown in Figure A.2 under the iPhone Developer Program's website:

http://developer.apple.com/iphone/program

Downloads

Xcode 4.2 for Lion
This is the complete Xcode developer
toolset for Mac, iPhone, and iPad. It
includes the Xcode IDE, iOS Simulator,
and all required tools and frameworks
for building Mac OS X and iOS apps.

Download Xcode 4

Posted Date: October 12, 2011
Build: 4D199
Included iOS SDK: iOS 5
Included Mac SDK: Mac OS X 10.7

Xcode 4.2 for Snow Leopard
Download ▶

Looking for Xcode 3? ▶

Posted Date: October 12, 2011
Build: 4C199
Included iOS SDK: iOS 5
Included Mac SDK: Mac OS X 10.6

iAd Producer 1.2
iAd Producer, combined with the
power of iOS 4 and its WebKit-based
browser, makes it easy for you to
create high-impact, motion-rich ads.

Download

iTunes 10.5.1 beta
This is a pre-release version of
iTunes 10.5.1 beta with iTunes Match
beta.

Important: iTunes Match beta is
currently available to developers in
the United States.

Posted: October 11, 2011

Downloads
📄 iTunes 10.5.1 beta with iTunes Match (Mac)
📄 iTunes 10.5.1 beta with iTunes Match (Windows)
📄 iTunes 10.5.1 beta with iTunes Match (Windows 64)
📄 iTunes 10.5.1 beta Release Notes

Figure A.1   Downloading Xcode on Apple Developer website.

Figure A.2   Registering a developer account.

Figure A.3   iOS SDK tools.

A.2   What Can You Create Using iOS SDK?

iOS is based on the UNIX and further built with all of the interfaces, tools, and resources needed to develop applications from your Intel-based Macintosh computer. Apple delivers most of its system interfaces in special packages called frameworks. A framework is a directory that contains a dynamic shared library and the resources.

iOS supports the development of two types of applications: Native and Web applications. Native applications appeared on the device's Home screen only and are installed directly on the device. Web applications use a combination of HTML techniques to implement interactive applications on the Internet.

A.2.1   What Tools Are in the iOS SDK?

After you have installed the iOS SDK, you will find there is a bunch of development applications under /Developer/Applications. Figure A.3 shows the icons of tools included in iOS SDK. The following paragraphs describe each included tool in the iOS SDK.

Xcode

Xcode provides the integrated development environment (IDE) for the whole project management. It is a powerful development program which includes tools for writing, debugging and compiling the source code. Xcode allows you to program different programming languages, including C, C++ and Objective-C. The main programming language to develop iPhone applications is Objective-C language. Similar to C language, Objective-C consists of header files and main coding files. Objective-C is slightly different from C language in syntax, pre-defined methods and naming of files. Unlike C and Java, it uses infix notation.

iOS Simulator

Xcode provides the iOS simulator which simulates the iPhone environment for helping the application run, test, and debug on the development computer locally. iOS simulator can simulate most application environments of iPhone, iPod touch and iPad which help you to do a careful testing before you deploy your program in a iOS device. However, the iOS simulator has its own limitations which cannot simulate the real-time features, such as GPS, digital compass, motion handling etc.

Interface Builder

Xcode provides an Interface Builder to create the layout of user interface by drag and drop interface creation and live preview for your new application.

Instruments

Instruments is a suite of performance analytics tools. It is used for monitoring and optimizing iPhone application performance in real time.

Dash Code

Dash Code is a simple script programming environment to create web applications, such as Javascript, CSS style sheet.

Quartz Composer

Quartz Composer is a node-based visual programming tool based on OpenGL, OpenCL, Core Image, Core Video, JavaScript, and other technologies for processing and rendering graphical data.

A.2.2 Apple Developer Center

Apple also provides useful resources on the official website for iPhone developers. Developers can access the iPhone Developer Center where they can get useful support for developing iOS applications. The following describes help provided by the iOS Developer Center.

Getting Started Videos & Documents

There are a number of videos and documents to help new developers to practice the iOS application development. The getting started videos and documents cover a large range of topics about the usage of tools and frameworks.

iOS Reference Library

iOS Developer Center provides developer news, technology releases, and technical information. Moreover, it provides a comprehensive technical specification of each framework. Developers can explore the reference library and learn a lot of useful hands-on programming skills and techniques. All documentations, such as API, guides and articles are carefully organized, so that the developer can find the information easily. Figure A.4 shows the iOS Reference library web page. You can access the iOS Developer Library at:

https://developer.apple.com/library/ios/navigation/index.html

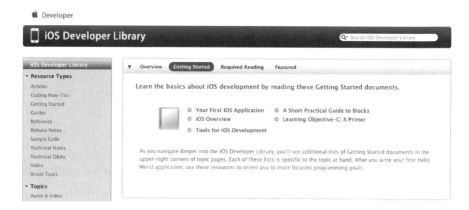

Figure A.4   iOS Developer Library.

Sample Code

iOS Developer Library provides a lot of sample code to study for implementing iOS applications. Each project in the sample code is buildable with executable source examples of how to accomplish a task for a specific technology. They show the correct sequence of calls and parameter data types to provide a generalized method for API use that developers can modify for their specific needs.

Apple Developer Forums

The iOS Developer Center provides the forum platform for users to post the topics and questions about iOS development. The developer can share their experiences about iOS application development on the forum. This is also an interactive platform between iPhone developers and Apple engineers to discuss iOS SDK experiences.

A.3   Limitations of iPhone Environment

iPhone application development is different from desktop application development. The following are the characteristics of the application environment in iPhone.

Only One Active Application

Because of the limited resources in iPhone, it only allows the running of one active application program at a time. In iOS5, there are five possible states for an application:

- ▶ Not running – Application has not been launched and nothing happens.

- ▶ Inactive – Application is running in the foreground but not receiving events.

- ▶ Active – Application is running in the foreground and receiving events.

- ▶ Background – Application is in the background and executing code.

- ▶ Suspend – Application is in the background but not executing code.

Access Control

iPhone has the strict restriction to limit users to access the iPhone operating system. The iPhone operating system is different from the other system such as Windows and Linux. Users can access and change file structure in the Windows and Linux system. iPhone OS provides an area called Sandbox for users to store the application which includes the programs, documents, or files used in the application. Users are only permitted to control those files in the sandbox.

Responsiveness

iPhone applications limit the response time in saving and controlling files. If users save the files over 5 seconds, then the process will be killed. This action is to guarantee the iPhone's applications run smoothly. Therefore, iPhone application developers should ensure that any data is saved within 5 seconds when quitting the application.

Display Size

The iPhone screen size is fixed as 480x320 pixels. Although iPhone's resolution is fine for a mobile phone, the resolution is still not better than a desktop or laptop. Therefore, iPhone application developers avoid putting unnecessary components in the application interface. The pithy interface makes the application more user-friendly.

A.4 Introduction to Xcode

Xcode provides a user-friendly interface for the developer to program. Like common software developing environment, Xcode includes compiler, interface builder, programming areas. Xcode was introduced in 2003 with version 1.0. It has been updated since then by introducing 2.0 with Mac OS X Tiger and Xcode 3.0 with Mac OS X Leopard. Finally, 4.2 was introduced, between releases of Mac OS X 10.6 Snow Leopard and Mac OS X 10.7 Lion.

With version 3.x and previous versions, Xcode comprised a few different applications: Xcode itself, Interface Builder and Organizer. Xcode is used to do the actual coding, create classes and even debug. Interface Builder, as the name implies, is for building the interfaces for iOS and Mac OS applications. The organizer is where you provision devices (for iOS development), archive code, and conduct performance tests.

Xcode 4 has made a significant change. Instead of three different applications, there is just one single interface for the entire application. They are all organized under Xcode 4 now.

When you click on the Xcode icon, it appears as a window as shown in Figure A.5 which provides four options:

▶ Create a New Xcode project – This creates a new application project and brings you to the coding environment.

▶ Connect to a repository – If you have existing projects and want to integrate them with Xcode's source control features, this brings you to connect to a GitHub or Subversion (SVN) repository and check-out or clone the project. With this being

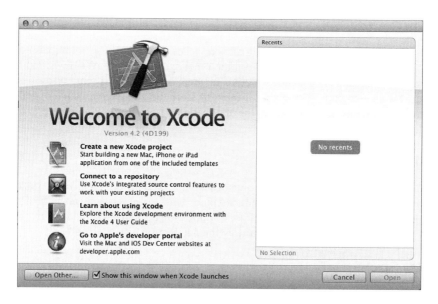

Figure A.5   Initial opening window of Xcode.

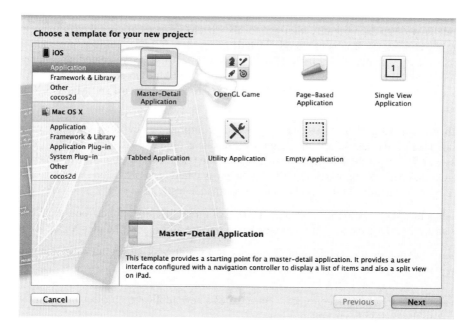

Figure A.6   Seven project templates in Xcode 4.

built in, it will make the process of code organization and collaboration significantly easier for teams of coders to work on the same project.

▶ Learn about using Xcode – This brings you to a user guide about the Xcode environment and shows how it operates.

▶ Go to Apple's developer portal – This links to the developer portal website.

A.4.1   Xcode Project Template

After you make the selection to create a new project, Xcode 4.2 prepares 7 application templates to start a new project as shown in Figure A.6.

▶ Master-Detail Application – This provides a master navigation controller which it uses to display the detailed list of items.

▶ OpenGL Game – This is for game development. It includes a subset of OpenGL 3D graphic API which is for embedded systems such as mobile phones, PDAs.

▶ Page-Based Application – This provides a page view controller.

▶ Single View Application – This gives only a view controller and a user interface template file with .nib extension. It is the most common template for the beginner developer.

▶ Tabbed Application – This starts with a tab bar and view controllers which has two .xib files for two different types of framework.

▶ Utility Application – There is a button to flip the main view to the flipside view. Also, there is a navigation bar for flipping back to main view.

▶ Empty Application – This provides an empty window-based template. This template provides a window to users. It can develop any type of window-based application.

A.4.2   Xcode Project Summary

After creating and saving your new project, Xcode 4 will present you with the summary of all the configuration options that you can set regarding your application. Figure A.7 shows a project summary in Xcode 4. If you need to change the type of application from iPad to iPhone/iPod touch or universal, you can do so quickly by just changing the drop-down option. Once you do select the drop down, you can see new options pertaining to your change appear.

Depending on your settings and previous uses of Xcode, your window might not appear to be the same as in the Figure A.7. But it should not have much difference about the basic layouts. You can always go back to the project summary if you click the root of groups and files in the navigation area.

Figure A.7   Project summary in Xcode 4.

A.5   Xcode Project Interface

We can divide the Xcode interface into five components: toolbar, navigation area, editor area, debug area and utility area. Figure A.8 shows the five components of Xcode environment for developing iPhone applications.

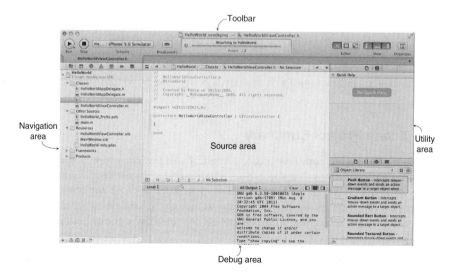

Figure A.8   Xcode environment for developing iPhone applications. (A full color version of this figure appears in the color plate section.)

A.5.1 Toolbar

The major function of the toolbar is to compile and run the application. You can simply click the play button to build the application or click the stop button to terminate the application.

On the left side of the Xcode toolbar as shown in Figure A.9, you can see a pull down menu with the current selection of simulator and the active devices (iOS devices connected to the computer). Most of the time, you test the application under the simulation mode. Unless you need to deploy it to a real iOS device, you need to choose the active iOS device.

Figure A.10 shows the Xcode toolbar and provides simple descriptions of each button. You can easily enable/disable the breakpoints placed in the code by clicking the breakpoint button in the toolbar. You can change the layout of the source area to see your history of code writing with a split view by clicking a button in the Editor button set (located on the right hand side of the toolbar). You can open/close the navigation area, debug area and utility area by clicking the button in the View button set. You can always launch the Organizer window to see detailed information of the available iOS devices, repositories of projects, edited project descriptions, archived projects and to look for help in documentation.

Figure A.9   Choosing the iOS simulator version.

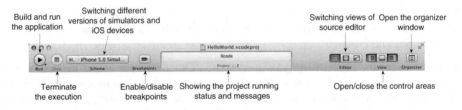

Figure A.10   Xcode toolbar.

A.5.2   Navigation Area

Navigation area provides you a view of group and files that manages the whole project. There are seven small navigator buttons located at top of the navigation area, 'Project Navigator,' 'Symbol Navigator,' 'Search Navigator,' 'Issue Navigator,' 'Debug Navigator,' 'Breakpoint Navigator,' and 'Log Navigator,' as shown in Figure A.11.

▶ Project Navigator – This provides a master view of all files for the project, such as, source code, compile sources, libraries etc. (We will talk about it later.)

▶ Symbol Navigator – This shows the hierarchical structure of object instances in the project in terms of class, method, variable etc.

▶ Search Navigator – This provides a search field to search any text in the project.

▶ Issue Navigator – This gives you the warning and error message after the compilation.

▶ Debug Navigator – This shows and monitors the debug instances when the project successfully in a run-mode.

▶ Breakpoint Navigator – This shows the existing breakpoints in the project.

▶ Log Navigator – This gives log files about any actions, e.g. build and debug performed in the project.

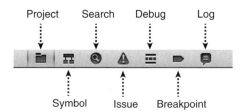

Figure A.11   Seven navigators in the navigation area.

Project Navigator

Users can find all information for the project, for example, source code, compile sources, libraries etc. in the Project Navigator. Figure A.12 shows different groups and files in the Project Navigator.

When you click onto the root, it shows the project summary. The first folder is named by the project name. It includes the five subfolders. They are Classes, Other Sources, Resources, Frameworks, and Products.

'Classes' stores the code which developers write. 'Other Sources' stores the source code which are not Objective-C classes. 'Resources' stores all the noncode files which used in the application, for example sound file, image files. 'Frameworks' stores the special library or framework such as UIKit Framework which used in application. 'Product' stores the application after compilation; it is in .app extension.

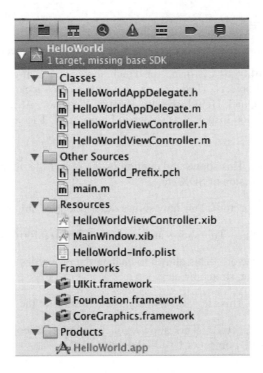

Figure A.12   Xcode Project Navigator.

A.5.3   Editor Area

Editor area provides a 'smart' code sheet for writing the source code of application. The 'smart' sheet includes many functions, for example, showing hints of arguments about user-defined or system-defined methods, providing selections of classes and methods, checking and warning the wrong syntax.

A.5.4   Debug Area

Debug area consists of two parts: breakpoint debug area (left) and output debug area (right). If breakpoints in the project exist, the breakpoint debug area can show all the 'freeze' values of local and global variables. The output debug area shows the console output of the application. Figure A.13 shows an example of output in debug area.

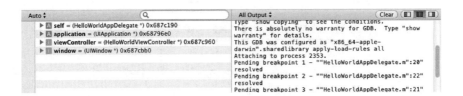

Figure A.13   Xcode debug area.

A.5.5   Utility Area

Utility area consists of two parts: Inspector (upper) and Library (lower).

The upper part of utility area is Inspector which shows and configures the properties of project, methods and objects. There are six inspector buttons located at the top of utility area, File Inspector, Quick Help Inspector, Identity Inspector, Attributes Inspector, Size Inspector and Connection Inspector as shown in Figure A.14. The last four inspectors only appear in UI elements.

  ▶ File Inspector – This shows and configures the object file information, such as file name, which version of SDK etc.

  ▶ Quick Help Inspector – This only shows descriptions of any object.

  ▶ Identity Inspector – This shows and configures the ID information of the UI object

  ▶ Attributes Inspector – This allows to change the properties of a UI object such as color, font, offset, stretching etc.

  ▶ Size Inspector – This allows you to configure the size of the UI object.

  ▶ Connection Inspector – This allows you to connect/disconnect the relationship among code-to-UI object or UI-to-UI objects.

The lower part of the utility area is Library which consists of four types of library, File Template Library, Code Snippet Library, Object Library and Media Library.

  ▶ File Template Library – This provides any templates in the form of file, such as class file, view file, data model file, property list file etc.

  ▶ Code Snippet Library – This provides any templates in the form of code snippet, such as, functions, structures, name spaces etc.

  ▶ Object Library – This provides any templates in the form of object, such as user interface objects, controller objects, view objects etc. It is a frequently used library to build the user interface of applications. Figure A.15 shows the object library.

  ▶ Media Library – This provides any templates in the form of media, such as music, movie, etc.

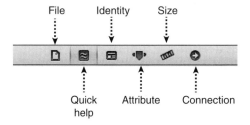

Figure A.14   Six inspectors in the utility area.

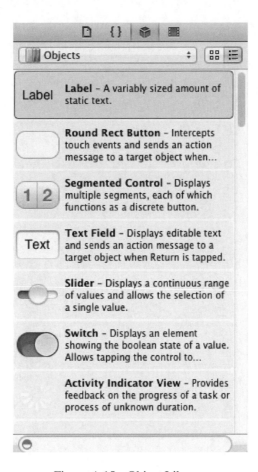

Figure A.15   Object Library.

Appendix Summary

After reading this chapter, you should know about the background and tools of iOS SDK. You have also learned the basic components of the Xcode development platform. The next appendix chapter introduces the Objective-C language and programs on some simple iPhone applications.

APPENDIX B

INTRODUCTION TO OBJECTIVE-C PROGRAMMING IN iPHONE

What is in this appendix chapter?

▶ Introducing the user interface of Xcode

▶ Creating a command line project in Xcode

▶ Introducing the syntax of Objective-C language

▶ Building iPhone applications in Xcode

This is the second appendix chapter to introduce basic programming techniques in iPhone. The language used in iPhone SDK is Objective-C. It is a superset of ANSI C and started out as just a C preprocessor and a library, but over time developed into a complete runtime system, allowing a high degree of dynamism and yielding large benefits. We assume that you have some programming experiences and understand what object-oriented programming (OOP) is. Similar to C language, Objective-C can consist of header files and main coding files. Objective-C is slightly different from C language in syntax, pre-defined methods and naming of files. After you read through this chapter, you will be able to write a small program in iPhone.

We first introduce the user interface of Xcode. Second, we briefly introduce new features of OOP in Objective-C language. Finally, we teach you to develop the user interface and code in iPhone. This chapter teaches Objective-C programming in iPhone by example. A small complete program will cover each new features of Objective-C language or iPhone. Practices make perfect. It would be better if you could learn through the new features by running each program. Also, this will help you to get familiar with the syntax of Objective-C.

Before You Start

As you begin to learn programming in iPhone, you should ensure you have installed Apple's Xcode tools. If you do not have Xcode in your Mac machine, please go back to Appendix A and get your Xcode ready.

B.1   Objective-C Program, HelloWorld

To begin, let's get start with the classic HelloWorld program which displays the text, 'Hello, world!.' Instead of using .c file extension, Objective-C uses .m file extension to store the source code. Program Appendix B.1 shows an Objective-C HelloWorld program:

Program Appendix B.1

```
// HelloWorld.m program example

#import <Foundation/Foundation.h>
int main (int argc, const char * argv[])
{
    @autoreleasepool {
        // insert code here...
```

```
      NSLog (@"Hello, World!");
    }
    return 0;
}
```

B.1.1   Using Xcode to Code and Compile Programs

Xcode is a powerful programming tool to create your Objective-C projects. If you have successfully installed iPhone SDK, you can find Xcode application in the following default directory /Developer/Applications. When you launch Xcode, please click the menu and select **File** − > **New Project** as shown in Figure B.1.

You choose **Application** in the lower-left pane of Mac OS X. In the upper-right pane, you should highlight the **Command Line Tool**, and click **next** as shown in Figure B.2.

Type **Hello World** in the Product Name field and press next as shown in Figure B.3. Then scroll down the **Type** menu. Finally, select **Foundation** as shown in Figure B.4. Click **Next**. A window appears as shown in Figure B.5.

You can choose the location of the project, select **Documents** folder in the left hand sidebar and click **Create**. A new window appears as shown in Figure B.6.

In the upper-right pane, select HelloWorld.m. You can see the Xcode project window automatically generate the code of Program Appendix B.1 into lower-right pane. You can compile and run by clicking the 'Run' button at the top right hand corner. If everything

Figure B.1   Creating a new project in Xcode.

Figure B.2   Creating a new project -> Selecting **Command Line Tool**.

Figure B.3   Type **HelloWorld** in the Product Name field.

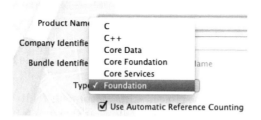

Figure B.4   Selecting **Foundation**.

goes smooth, an Xcode Debugger Console window displayed at the bottom right hand side window should appear as in Figure B.7.

Maybe, you are still panic or you are quite familiar of the code because it looks like C language. However, in some ways, it is different from C language. Let's go through **Program Appendix B.1** line by line to study the differences.

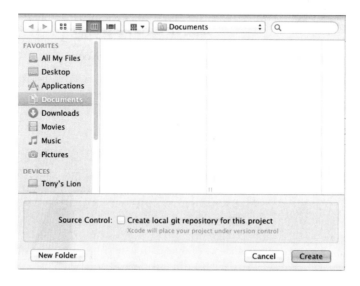

Figure B.5   Save a new project in Xcode.

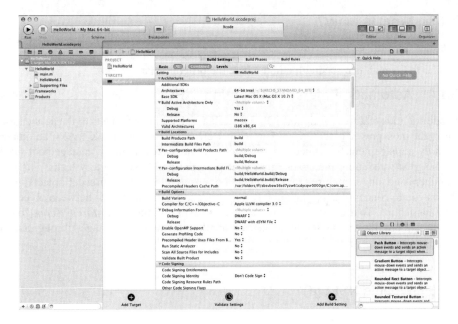

Figure B.6   HelloWorld project window.

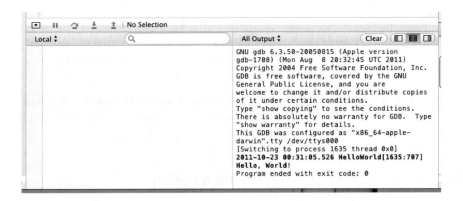

Figure B.7   HelloWorld program runs in a console environment.

B.1.2   What is #import?

In C, you can use #include a header file for some definitions. However, in Objective-C, you need to use #import, like this:

#import is used to include a header file that informs compiler you are compiling Objective-C programs. #import could avoid recursively including redundancy files, whereas C use #ifdef to avoid. <Foundation/Foundation.h> is a header file of Foundation framework that handles functions such as data input, user interface, communication, etc.

B.1.3   What is Main?

```
int main (int argc, const char * argv[])
```

Similar to C, main is a special method where the program starts to run.

B.1.4   Automatic Reference Counting (ARC)

Xcode 4.2 adds features to support iOS 5 as well as other enhancements to the application; one of the most powerful features is Automatic Reference Counting (ARC). ARC automates memory management for Objective-C objects, and it makes memory management much easier, greatly reducing the chance that your program will have memory leaks. The programmer is no longer required to think about retain and release operations of objects; ARC allows you to concentrate on the interesting code, and the relationships between objects in your application. To use ARC with your application, simply start a @*autoreleasepool* block in the main program, and ARC will take care of all of the memory managed throughout the whole program.

```
@autoreleasepool { ... }
```

B.1.5   What is NSLog()?

```
NSLog (@"Hello, World!");
```

Instead of using printf(), NSLog() displays the string with time, date stamps and the newline $('\backslash n')$. @ sign followed by a pair of double quotes indicating that the following are a series of strings of character.

```
return 0;
```

The above statement in the main terminates execution and returns 0 to the main method. The usage of return in here is same as C.

Tremendous! You understand the above Objective-C program thoroughly. You probably understand how the infix syntax makes Objective-C different from traditional C language. Because of the limited space of the content, let's move to object-oriented programming (OOP) which contains more new features and syntax of Objective-C.

B.2   Object-Oriented Programming (OOP)

OOP in Objective-C has similar features and concepts of other programming languages, such as C++, C# and Java. You probably understand the concept of a **class**, an **object**, a **instance**, a **message**, a **method** and **method dispatcher** from other programming languages.

Let's think about an example to define a class of mobile device. You can name the class as Mobile. Maybe, your mobile phone could only store a phone number and a message. You could create an instance of Mobile, such as:

```
myMobile = [Mobile new];    get a new mobile
```

You may have some functions in your mobile device. You could create methods for your mobile device:

```
[myMobile storeNumber]        store telephone number
[myMobile getNumber]          display a telephone number
```

```
[myMobile sendMessage]          send a message
[myMobile receiveMessage]       receive a message
[myMobile getMessage]           get a message
```

Now, we could create an actual class in Objective-C and learn how to work with an instance of the Mobile class. **Program Appendix B.2** shows the code of a Mobile program (Figure B.8).

Program Appendix B.2

```objectivec
// Mobile.m program example

#import <Foundation/Foundation.h>

//-- @interface section --

@interface Mobile : NSObject
{
    NSString* ownerName;
    long number;
    NSString* message;
}
-(void) setOwnerName:(NSString*) name;
-(NSString*) getOwnerName;
-(void) storeNumber:(long)num;
-(long) getNumber;
-(void) sendMessage;
-(void) receiveMessage:(NSString*)text;
-(NSString*) getMessage;
@end // Mobile

//-- @implementation section --

@implementation Mobile

-(void) setOwnerName:(NSString*) name
{
    ownerName = name;
}
-(NSString*) getOwnerName
{
    return ownerName;
}
-(void) storeNumber:(long)num
{
    number=num;
}
-(long) getNumber
{
    return number;
}
-(void) sendMessage
```

```
{
     NSLog(@"Sending Message: Hello");
}
-(void) receiveMessage:(NSString*)text
{
     message=text;
}
-(NSString*) getMessage
{
     return message;
}
@end // Mobile

//-- main program section --

int main (int argc, const char * argv[]) {
    @autoreleasepool {
         Mobile *myMobile;

         // Create an instance of a Mobile
         myMobile = [Mobile new];

         //Set owner of the phone
         [myMobile setOwnerName:@"Eddie"];

         //Display the owner
         NSLog(@"The owner is %@",[myMobile getOwnerName]);

         //Receive a message
         [myMobile receiveMessage:@"Hello"];

         //Display a message
         NSLog(@"The received message is %@",[myMobile getMessage]);

         //Store a nmuber
         [myMobile storeNumber:27664547];

         //Display a mobile number
         NSLog(@"Mobile Number is %li",[myMobile getNumber]);
    }
    return 0;
}
```

B.2.1   Infix Notation

```
myMobile = [Mobile new];
[myMobile setOwnerName:@"Eddie"];
```

The above statement creates an instance of Mobile object and assigns to the myMobile pointer. Objective-C uses infix syntax. The name of the method and its arguments are

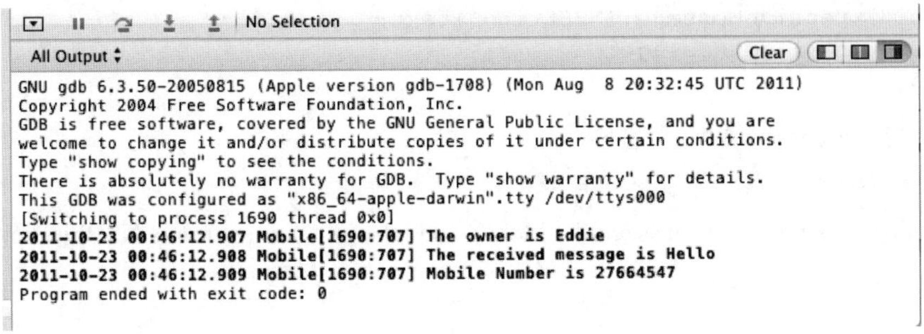

```
All Output ▾                                                    Clear
GNU gdb 6.3.50-20050815 (Apple version gdb-1708) (Mon Aug  8 20:32:45 UTC 2011)
Copyright 2004 Free Software Foundation, Inc.
GDB is free software, covered by the GNU General Public License, and you are
welcome to change it and/or distribute copies of it under certain conditions.
Type "show copying" to see the conditions.
There is absolutely no warranty for GDB.  Type "show warranty" for details.
This GDB was configured as "x86_64-apple-darwin".tty /dev/ttys000
[Switching to process 1690 thread 0x0]
2011-10-23 00:46:12.907 Mobile[1690:707] The owner is Eddie
2011-10-23 00:46:12.908 Mobile[1690:707] The received message is Hello
2011-10-23 00:46:12.909 Mobile[1690:707] Mobile Number is 27664547
Program ended with exit code: 0
```

Figure B.8   Mobile program output.

all intertwined. The next statement calls the setter method of property OwnerName with parameter.

There are two new types of syntax used in Objective-C, **Interface** and **Implementation**.

An **Interface** is usually in header files of Objective-C program, which provides the interface of a class. However, an **Interface** does not provide concrete method and only provides an abstract layout of a class, such as its data components and its method. An **Implementation** has the actual code that makes these method defined in an **Interface** workable. Usually, a class can be divided into three sections, @interface section, @implementation section and program section.

B.2.2   The @Interface Section

This part of the code starts with @interface and the class name, and has @end, just like the interface. All methods must appear between these two statements. The @interface section usually could be separated in the header file. Class names usually begin with an uppercase letter by convention. Method and variable names should be in lowercase letters.

Normally, there are three parts in @Interface section. First, you should define the class name (in tradition, first letter should be capital for the class name) and where it inherits. Second, you should declare arguments or parameters that describe the data members of a class. These parameters are called the instance variables. Finally, you need to think about methods, which provide functions in a class. The below code is the @interface for the Mobile class:

```objc
@interface Mobile : NSObject
{
    NSString* ownerName;
    long number;
    NSString* message;
}
-(void) setOwnerName:(NSString*) name;
-(NSString*) getOwnerName;
-(void) storeNumber:(long)num;
-(long) getNumber;
-(void) sendMessage;
```

```
-(void) receiveMessage:(NSString*)text;
-(NSString*) getMessage;
@end // Mobile
```

Let's look at the first line of code.

```
@interface Mobile : NSObject
```

This line declares a new class, called Mobile and inherits features from a NSObject class.

```
NSString* ownerName;
long number;
NSString* message;
```

Similar to C++ programming, the above lines declare three instance variables.

There is some new syntax in method declarations. Let's look at the following line:

```
-(void) receiveMessage:(NSString*)text;
```

Objective-C uses the leading dash to start to declare an Objective-C method. A single dash before a method name means it's an instance method. A plus before a method name means it's a class method. The parentheses after the dash is the return type of the method. Similar to C, we use void to indicate that there is no return value.

As mentioned, Objective-C uses infix notation to receive arguments. The colon after receiveMessage method name indicates to receive input argument. If we have more than one argument, a colon is used to separate each argument. For example, we need to receive a message with a text and an incoming mobile phone. We could modify the above line as follows:

```
-(void) receiveMessage:(NSString*)text: (long)number;
```

We do not go through all the details in the above codes because we assume you to know most of the OOP concepts in C++ or Java programming.

That's great. We have successfully completed the interface section. Let's go to the implementation section.

B.2.3   The @Implementation Section

The @implementation section defines the actual methods concretely from the @interface section. The below code is the @implementation for the Mobile class:

```
@implementation Mobile

-(void) setOwnerName:(NSString*) name
{
     ownerName = name;
}
-(NSString*) getOwnerName
{
     return ownerName;
}
-(void) storeNumber:(long)num
```

```
{
     number=num;
}
-(long) getNumber
{
     return number;
}
-(void) sendMessage
{
     NSLog(@"Sending Message: Hello");
}
-(void) receiveMessage:(NSString*)text
{
     message=text;
}
-(NSString*) getMessage
{
     return message;
}
@end // Mobile
```

The first line of each method is the same as in @interface. Each method is declared in @interface which corresponds to the method declaration in @implementation. Instead of ending with semicolon, it follows with the functional code inside a set of curly braces. You can code without following the list order in @interface section. You even could rename different input arguments that differ from @interface section. By convention, it is better to keep it the same. The remaining part in @implementation section is quite similar to C++ programming language.

B.2.4   The Program Section

We usually solve problem in the program section. In OOP, we instance objects in the program section. The below code is the program section for the Mobile class:

```
int main (int argc, const char * argv[]) {
    @autoreleasepool {
         Mobile *myMobile;

         // Create an instance of a Mobile
         myMobile = [Mobile new];

         //Set owner of the phone
         [myMobile setOwnerName:@"Eddie"];

         //Display the owner
         NSLog(@"The owner is %@",[myMobile getOwnerName]);

         //Receive a message
         [myMobile receiveMessage:@"Hello"];
```

```
                //Display a message
                NSLog(@"The received message is %@",[myMobile getMessage]);

                //Store a nmuber
                [myMobile storeNumber:27664547];

                //Display a mobile number
                NSLog(@"Mobile Number is %li",[myMobile getNumber]);
    }
       return 0;
}
```

We create an instance of a mobile object in infix notation in the line below:

```
myMobile = [Mobile new];
```

Inside the setOwnerName: the method is shown below:

```
[myMobile setOwnerName:@"Eddie"];
```

We need to pass a pointer instance of NSString class in myMobile, however, 'Eddie' is not a pointer instance of NSString class. We use an anonymous casting, (NSString*) to create an anonymous pointer instance of 'Eddie.' If we define setOwnerName: method in C++ language, it will be:

```
myMobile.setOwnerName((NSString*) ''Eddie");
```

As mentioned of ARC before, the system will keep track on the objects memory usage and release the memory occupied by the unused objects automatically by using ARC. There is no garbage collection is required as in Xcode 4.1 before.

B.2.5   @property, @synthesize and @dynamic

Now, you have learnt how to define a class, call a method and pass the data into a method in Objective-C. Let's continue to talk some new features in Objective-C 2.0. In OOP, you probably know a concept of 'encapsulation.' Very often, we use setter and getter method to achieve information hiding to set and get parameters. Objective-C 2.0 provides auto setter and getter method (or, mutator and accessor). @property is declared in @interface section which tells the compiler to use auto setter and getter methods. @synthesize is declared in @implementation section which tells the compiler to synthesize auto setter and getter methods. **Program Appendix B.3** shows the modified code of a Mobile program.

Program Appendix B.3

```
// Modified Mobile.m program example

#import <Foundation/Foundation.h>

//-- @interface section --
@interface Mobile : NSObject
{
    NSString* ownerName;
    long number;
```

```
        NSString* message;
}

@property long number;
@property (copy) NSString* ownerName;
@property (copy) NSString* message;
@end // Mobile

//-- @implementation section --
@implementation Mobile
@synthesize ownerName, number, message;

@end // Mobile

//-- main program section --
int main (int argc, const char * argv[]) {
    @autoreleasepool {
            Mobile *myMobile;

            // Create an instance of a Mobile
            myMobile = [Mobile new];

            //Set owner of the phone
            [myMobile setOwnerName:@"Eddie"];

            //Display the owner
            NSLog(@"The owner is %@",[myMobile getOwnerName]);

            //Receive a message
            [myMobile receiveMessage:@"Hello"];

            //Display a message
            NSLog(@"The received message is %@",[myMobile getMessage]);

            //Store a nmuber
            [myMobile storeNumber:27664547];

            //Display a mobile number
            NSLog(@"Mobile Number is %li",[myMobile getNumber]);
    }
    return 0;
}
```

B.2.6   @property in the @interface Section

The @property is an Objective-C directive which declares the property. With this new feature, number of codes are reduced and simplified to define the getter and setter. The @property shrinks @interface section to declare attributes of a class. Let's see the in the line below how to define attributes in the @interface section:

```
@interface Mobile : NSObject
{
    NSString* ownerName;
    long number;
    NSString* message;
}
@property long number;
@property (copy) NSString* ownerName;
@property (copy) NSString* message;
@end // Mobile
```

Instead of getter and setter declaration, ownerName, number and message now become attributes of a class. @property long number; means that the objects of the class Mobile have an attribute, of type long, called number. However, @property does not take extra arguments for setter and getter. The name of the attribute is not necessarily the same as the instance variable.

Usually, the @property keyword is followed by parentheses. The value defined in the parentheses is the properties of an attribute. Assign, copy, retain, atomic, nonatomic, readwrite, readonly are the common properties of an attribute.

assign is to have a simple assignment of a setter method. This is the default.

copy is to make a copy of an attribute. Making a copy prevents the value from changing unexpectedly.

retain is to retain or reference an attribute. These mean that mutation of an object's attributes will be maintained behind its back.

atomic is to make sure an attribute always returns changes from the getter and setter, regardless of setter activity on any other thread. This is the default. For example, if thread A is in the middle of the getter while thread B calls the setter, an autoreleased object most likely will be returned to the caller in A. atomic is set by default.

nonatomic is to ignore the multi-thread changes by setter of an attribute. Thus, nonatomic is considerably faster than atomic. The resource-constraint environment like iPhone is usually recommended to use it.

readwrite is to allow reading and modifying an attribute. This is the default. Readonly is only to allow reading an attribute.

After we explain the properties of an attribute, we look at the meaning of the line below:

```
@property (copy) NSString* ownerName;
```

This means that the objects of the class Mobile have an attribute, of type NSString*, called ownerName with a copy property.

B.2.7   @synthesize in the @implementation Section

The @synthesize directive automatically generates the setters and getters. The compiler will synthesize the setter and/or getter methods for the property if they don't already exist within the @implementation block.

@synthesize ownerName, number, message; shrinks and generates accessor methods of these three attributes. ownerName, number, message are synthesized. Note that @synthesize can only be used in an instance variable.

B.2.8 @dynamic in the @implementation Section

The @dynamic keyword tells the compiler that you will fulfill the API contract implied by a property either by providing method implementations directly or at runtime using other mechanisms such as dynamic loading of code or dynamic method resolution. For example, you can tailor-make your own accessor of message in the @implementation section like so:

```
@dynamic message;

-(NSString*) message
{
    return message;
}
```

B.2.9 Dot Notation

In Objective-C, the dot notation is used to access object attributes. We can simply assign or retrieve value from a private variable in an Objective-C class just like a public variable defined in Java, C++ and C#. We can use dot notation in getter method like so:

```
NSLog(@"The owner is %@",myMobile.ownerName);
```

Or it is equivalent to

```
NSLog(@"The owner is %@",[myMobile ownerName]);
```

We can also use dot notation in setter method like so,

```
myMobile.number=27664547;
```

B.2.10 Category

Category is a new feature in Objective-C that adds new methods to existing classes without subclassing. It is very convenient to use category to add method to an Objective-C existing class, like, NSString. You do not need to subclass from it. For example, you want to add a method to determine whether a string contains a query of a string and ignore the case issue. **Program Appendix B.4** shows the example of using category.

Program Appendix B.4

```
//NSString-Has.h
@interface NSString (Has)
- (BOOL) hasThisString:(NSString*)s;
@end //Has


//NSString-Has.m
#import <Foundation/Foundation.h>
#import ''NSString-Has.h"
```

```
@implementation NSString (Has)
- (BOOL) hasThisString:(NSString*)s
{
    NSRange range = [self rangeOfString: s
            options:NSCaseInsensitiveSearch];
    if (range.location!=NSNotFound)
        return YES;
    else
        return NO;
}
@end //Has

int main (int argc, const char * argv[]) {
    @autoreleasepool {
        NSString* string1 = @"Objective-C is the best";
        NSString* string2 = @"C programming is easy to learn";
        NSString* string3 = @"Java is the BEST";
        NSString* query = @"best";

        if ([string1 hasThisString:query] )
            NSLog (@"String 1 contain \"%@\" string.",query);
        if ([string2 hasThisString:query] )
            NSLog (@"String 2 contain \"%@\" string.",query);
        if ([string3 hasThisString:query] )
            NSLog (@"String 3 contain \"%@\" string.",query);
    }
    return 0;
}
```

Program Appendix B.4 Output of a NSString-Has category class

Let's pull this apart piece by piece. First, you need to add a new name in parentheses in an existing class, NSString, just looks like this:

```
@interface NSString (Has)
- (BOOL) hasThisString:(NSString*)s;
@end //Has
```

This means that the category is called Has and it adds hasThisString: methods to NSString in the @interface section. We split the entire @interface section in a separated file, 'NSString-Has.h.'

Referring to the @interface section, we define the concrete methods in the @implementation section as follows:

```
#import ''NSString-Has.h"

@implementation NSString (Has)

- (BOOL) hasThisString:(NSString*)s
{
    NSRange range = [self rangeOfString: s options:
```

```
        NSCaseInsensitiveSearch];
        if (range.location!=NSNotFound)
            return YES;
        else
            return NO;
}
```

`@end //Has`

hasThisString method receive a NSString* argument and returns a Boolean object. Inside hasThisString: method, self is used to represent the original object itself. In C++ and Java programming, we use this to the same purpose. rangeOfString method receives two arguments, one is the string variable to be searched and one is a case-sensitive variable, NSCaseInsensitiveSearch. rangeOfString method return NSRange type of an object. NSNotFound is a value that indicates that an item requested couldn't be found or doesn't exist. If range.location is not equal to NSNOTFound, we return YES, otherwise, we turn NO. Let's see how we use our new method in NSString:

```
int main (int argc, const char * argv[]) {
    @autoreleasepool {
            NSString* string1 = @"Objective-C is the best";
            NSString* string2 = @"C programming is easy to learn";
            NSString* string3 = @"Java is the BEST";
            NSString* query = @"best";

            if ([string1 hasThisString:query] )
                NSLog (@"String 1 contain \"%@\" string.",query);
            if ([string2 hasThisString:query] )
                NSLog (@"String 2 contain \"%@\" string.",query);
            if ([string3 hasThisString:query] )
                NSLog (@"String 3 contain \"%@\" string.",query);
    }
    return 0;
}
```

Figure B.9 shows the output of the result. The result clearly shows that String1 and String3 contains a query. We have successfully created a new method, hasThisString into existing

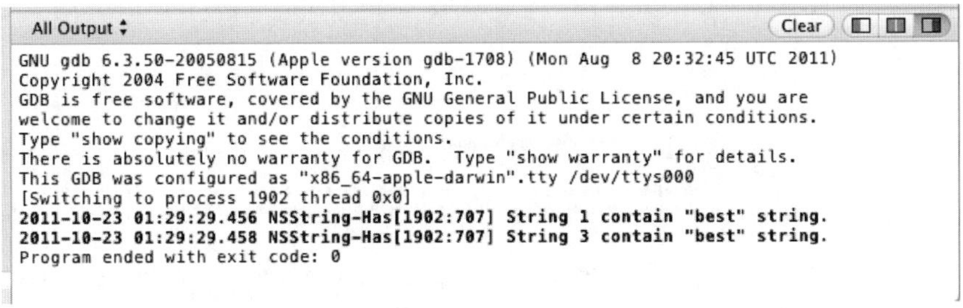

Figure B.9   NSString-Has category output.

built-up NSString class. There is no limit to how many categories may be added into the class. However, category cannot add instance variables to a class.

When you run this program, you will probably find out how powerful category is. There is no need to subclass a class and it just works. You may also use category features to override existing methods in a class.

So far, we have covered some new Objective-C features that may frequently be used in iPhone programming. Let's get started to program your iPhone application now.

B.3   HelloWorld iPhone Application

Probably, you are quite familiar with Xcode and Objective-C programming having gone through the previous sections. Now, open a new project in Xcode. Instead of choosing Mac OS X application, you should should choose **Application** in iPhone OS pane as shown in Figure B.10.

You choose **Single View Application** from a number of project templates in the upper right pane. **Single View Application** is the simplest template and easy-to-learn for your few iPhone applications. After you have selected **Single View Application**, a save box appears as in Figure B.11. This time, you should save the project template, HelloWorld in **Desktop** location.

The Xcode project window appears as usual, just as in Figure B.12. But this time, the right upper pane has a lot of different files. You could click and open each folder in the left pane. Let's discuss one by one from the top to the bottom of the left pane.

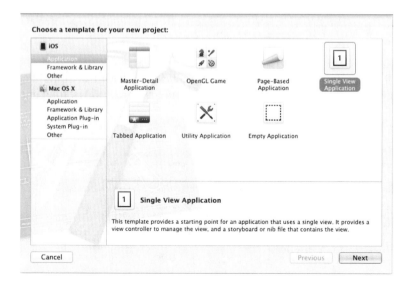

Figure B.10   Selecting **Single View Application** in Xcode.

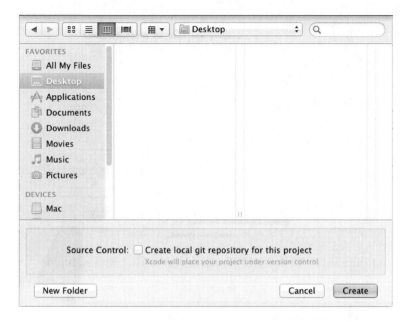

Figure B.11   Save your project, HelloWorld in **Desktop** location.

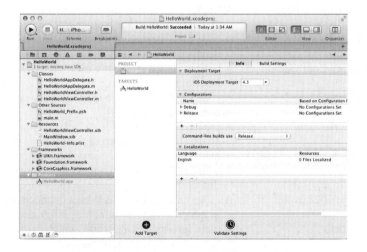

Figure B.12   Xcode Windows for HelloWorld Project.

B.3.1   Using Interface Builder

If you double click HelloWorldViewController.xib in Xcode window, Interface Builder window will appear as below:

The leftmost window shown in Figure B.13 is the **Library** window. **Library** window contains Tools menu. Similar to other programming tools, such as Microsoft Visual C++,

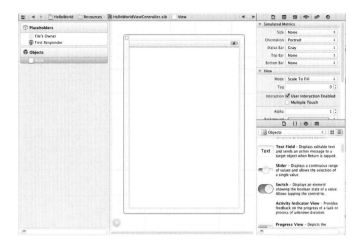

Figure B.13   HelloWorldViewController.xib in Interface Builder.

Table B.1   All files and their usage in HelloWorld program.

Files	Description
HelloWorldAppDelegate.h	A header file for creating a program window and an instance of interface builder and view controller
HelloWorldAppDelegate.m	A code file for creating a program window and an instance of interface builder and view controller
HelloWorldViewController.h	A header file for creating a user interface and handling user-driven event
HelloWorldViewController.m	A code file for creating a user interface and handling user-driven event
HelloWorld_Prefix.pch	A precompiled header file, which contains a list of header files from external frameworks that are used by our project
main.m	A main method. You do not need to edit or change this file
HelloWorldViewController.xib	An Interface Builder file, which connects with graphics design and graphics programming
MainWindow.xib	A main Interface Builder file, which loads HelloWorldViewController.xib
HelloWorld-Info.plist	An XML property list, which contains a lot of coding properties. Especially, it contains properties parameter of .xib file
*.framework	Frameworks are a special kind of library, which contains code, image and sound files. UIKit framework links to UI, Foundation framework links to NS and CoreGraphics links to graphics variable parameter

JBuilder and Ellipse, we can easily create the user interface by dragging and dropping the user interface elements from **Library** window to the **View** panel.

The middle window HelloWorldViewController.xib contains three icons, File's Owner, First Responder and View icons.

File's Owner represents the object that loaded the nib file from disk and owns this copy of the nib file. Imagine you have a nib file that contains the layout for your document and you also have an object as a controller for each of your documents that gets your data into the document. The File's Owner of the nib is the object that makes communication possible between this new nib and other parts of the application. If your application has just the one nib file, you really don't need to worry about File's Owner.

First Responder icon represents an object of the user interface control element that has the user's focus. So if you click on a user interface control, the nib sets that clicked user interface control as First Responder. The First Responder icon helps you to set properties of control or view without coding to determine which control or view that might be.

View icon represents the object instance of the UIView class. **View** window is the area that user can see and interact with user interface (UI) elements. In a complex application, you could have a multiple view of user interface. The rightmost **View Attributes** window helps to define the properties of UI elements in the **View** window.

B.3.2   Creating User Interface by Click-dragging Processes

You can add new objects in your iPhone interface window by click-dragging UI components from the **Library** window to the **View** window. For example, you can add a Label by clicking a **Label** in **Library** window and dragging it to **View** window. **Label** is a static text that displays information to the user. Figure B.14(a) shows the result of adding a new label in the **View** window.

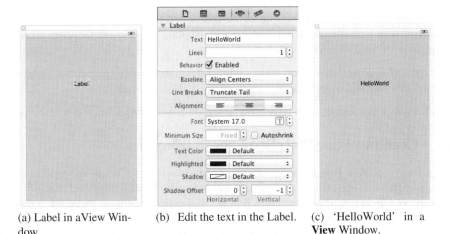

(a) Label in a View Window.

(b) Edit the text in the Label.

(c) 'HelloWorld' in a **View** Window.

Figure B.14   Click-dragging processes.

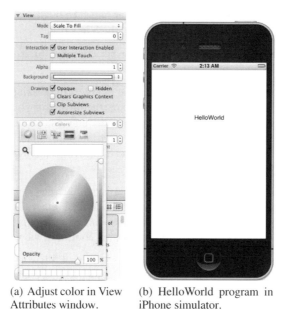

    (a) Adjust color in View   (b) HelloWorld program in
    Attributes window.   iPhone simulator.

Figure B.15   View Attributes.

You just need to double click the label and rename 'Label' to 'HelloWorld,' or you can use **Label Attributes** window to change the text field as shown in Figure B.14(b). You go back to **View** window and adjust the size of the label. Figure B.14(c) shows a label with the text of 'HelloWorld.' You can also change the background of the **View** window from grey color to white color by first clicking the background of **View** window and double clicking the background box in the **View Attributes** window. Figure B.15 shows how to adjust the background color in palette.

Finally, you select **Save** from the **File** menu and switch back to Xcode. Just simply click **Build and Run** button in the menu bar. Finally, Xcode compiles and launches an iPhone simulator with 'HelloWorld' displaying in the screen.

Is it magic? You do not need to touch any codes and you successfully launch the HelloWorld program in the iPhone simulator.

Let's keep this momentum and move a bit faster toward creating iPhone applications.

B.4   Creating Your Web Browser in iPhone

In this part, you know how to create your own web browser in iPhone. We need to decide how to implement the code in an appropriate way, which maintains the reusability and clearness of the code. Basically, most iPhone applications were designed by a concept called Model-View-Controller (MVC), which is a programming architectural model used in software engineering. Model represents classes held and processes the core data. View represents how the user interface interacts with user. Controller is between model and view, which handles event-driven data from view and passes the event-driven data to model.

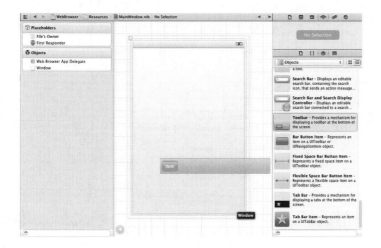

Figure B.16   Drag and drop the 'Toolbar' into window.

MVC model helps to distinguish these three types of codes and makes the code clearly separated. You will learn how to code some classes for web browser application (model), create a user interface of web browser through Interface Builder (view), and handle the user-driven events. (Controller) It is good to follow MVC model because it ensures the reusability and clarity of your code.

Let's get started with a new project in Xcode. For simplicity's sake, this time we also choose **Single View Application**. We save the project with the name WebBrowser. Let's create the user interface first. Double click MainWindow.xib in **Resources** folder. Now, a set of **Interface Builder** windows should appear.

In the **Object Library** window, you scroll down and find a tool item called 'Toolbar.' Drag and drop the 'Toolbar' into **Window** view as shown in Figure B.16.

You also drag and drop three extra **Bar Button** items to **Window** view just like Figure B.17(a). You double click the first button, name it as 'Back' and also double click the second button, name it 'Forward.' Click the third button. Go to Attribute Window -> Select First Icon -> Scroll down the pull-down menu in the Identifier field -> Choose Refresh. Similarly, choose Stop for the fourth button. Then, you will get the user interface as shown in Figure B.17(c).

Finally, you find the **WebView** item and drag it into **Window** as shown in Figure B.18(b). You have finished the **View** part. Let's go back to Xcode and and see the WebBrowserAppDelegate.h in the **Controller** part.

Program Appendix B.5a WebBrowserAppDelegate.h

```
#import <UIKit/UIKit.h>

@interface WebBrowserAppDelegate : NSObject
        <UIApplicationDelegate> {
```

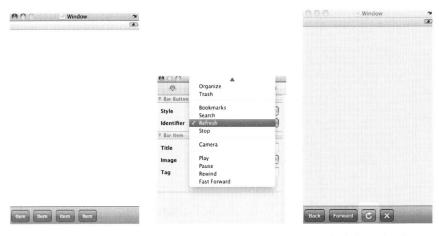

(a) Four new buttons in window.

(b) Select 'Refresh' in Attribute window

(c) Final view of toolbar

Figure B.17   Toolbar in window.

```
    UIWindow *window;
    UIWebView *webView;
}

@property (nonatomic, retain) IBOutlet UIWindow *window;
@property (nonatomic, retain) IBOutlet UIWebView *webView;

@end
```

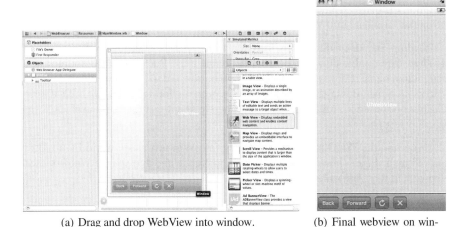

(a) Drag and drop WebView into window.

(b) Final webview on window.

Figure B.18   WebView in window.

The UIWindow class defines objects (known as windows) that manage and coordinate the windows in application displays on the screen. The UIWebView class defines objects that embed web content in your application. The IBOutlet constant is used to qualify an instance-variable declaration so that Interface Builder can synchronize the display and connection of outlets with Xcode.

Program Appendix B.5b WebBrowserAppDelegate.m

```
#import ''WebBrowserAppDelegate.h"

@implementation WebBrowserAppDelegate
@synthesize window;
@synthesize webView;

- (void)applicationDidFinishLaunching:(UIApplication *)application
{
    [webView loadRequest:[NSURLRequest requestWithURL:
    [NSURL URLWithString:@"http://maps.google.com/"]]];
    [window makeKeyAndVisible];
}

- (void)dealloc {
[webView release];
    [window release];
    [super dealloc];
}
@end
```

After we go through the header file, let's look at the code file to see how to load a URL string into a web browser. applicationDidFinishLaunching: method kicks off the application. This method operates automatically whenever the application launches. loadRequest method is used to load the web content inside the instance of UIWebView by passing an argument of the instance of NSURLRequest class.

NSURLRequest object represents a URL load request in a manner independent of protocol and URL scheme. requestWithURL method creates and returns a URL request for a specified URL with default cache policy and timeout value.

NSURL object provides a way to manipulate URLs and the resources they reference. URL-WithString method creates and returns an NSURL object initialized with a provided string. We pass a string of http://maps.google.com into URLWithString method.

After we define the **Controller** part, we need to link it with the **View** part. We can link them together by the **Interface Builder**. We double click the MainWindow.xib and click the Assistant editor button at the top of the window, then select WebBrowserAppDelegate.h in the lower part of the editor.

You choose webView and drag into UIWebView as shown in Figure B.19(a) which links from your **Controller** code to the **View** interface. Now, you click blue UIWebView and right click in the interface builder and a dialog will pop up, choose goBack in **Web View Connections** window and drag into 'Back' button in Figure B.19(b). Similarly, you drag go-Foward, reload and stop reloading into 'Forward,' 'Refresh,' and 'Stop' button respectively.

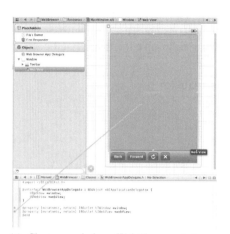

(a) Choose and drag WebView and drop into UIWebView.

(b) Drag goBack to 'Back' button.

Figure B.19   myView and UIWebView.

Figure B.20(a) shows the filled Received Actions. UIWebView class has declared these received actions automatically. Go to **Interface Builder** File menu, save the **View**.

Now, you go back to Xcode and click **Build and Run** button to run iPhone simulator. If you follow the steps, everything should be fine. Figure B.20(b) shows the result of

(a) Filled Received Actions in **Web View Connections** window.

(b) Web browser in iPhone simulator.

Figure B.20   View Connections.

web browser. Congratulations! You have successfully written your web browser in iPhone. You may feel weird that you do not touch main.m and the program just runs smoothly. Usually, we do not code in main.m because it is the **Model** part. You should see how simple and clear it is when we apply the MVC model to program.

B.5   Creating a Simple Map Application

Before we dive into the navigation programming and theories, let's do some more warm-up programming exercises. Google Maps provides a lot of APIs in iPhone. MapKit framework has been recently released in iPhone SDK and provides an interface for embedding maps directly into your own windows and views. This framework uses Google services to provide map data. Use of specific classes of this framework (and their associated interfaces) also provides support for annotating the map and for performing reverse-geocoding lookups to determine placemark information for a given map coordinates. The MapKit framework binds you to the Google Maps/Google Earth API terms of service as in the following link: http://code.google.com/apis/maps/iphone/terms.html.

As we do not want to overload you with too many new concepts, we will just work with some simple map applications in this chapter. As usual, we start a new project in Xcode and choose to use **Single View Applications**. You can save the file and name it, SimpleMap as Figure B.21. We double click SimpleMapViewerController.xib.

Interface Builder will appear as usual. Again, if you could not see any Interface Builder pop-up, just go to File Menu to open **View**, **Utilities**, **Show Object Library**.

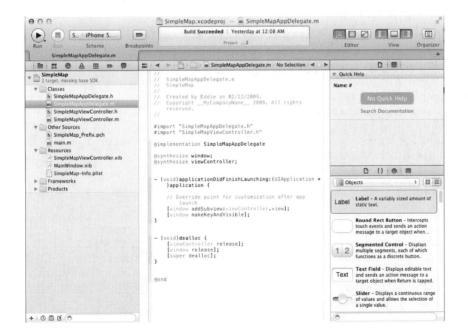

Figure B.21   Xcode windows for simple map.

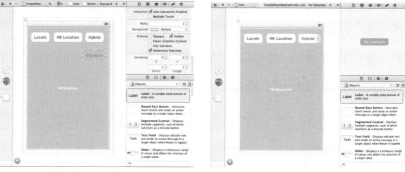

(a) Hide 'Standard' button.   (b) User interface of SimpleMap application.

Figure B.22   Edit application interface.

You should now drag a **MapView** item from Library windows to **View** window. Please also add four buttons at the top of **View** window and name them, 'Locate,' 'HK Location,' 'Hybrid' and 'Standard' respectively. For 'Standard' button, you initially set it as hidden in Button Attribute window as shown in Figure B.22(a). Finally, use 'Standard' button to cover 'Hybrid' button, as in Figure B.22(b).

MapKit framework is required for this application. We go back to Xcode and add an existing framework called MapKit framework into the Framework folder as shown in Figure B.23. If you cannot find it, please download an updated version of iPhone SDK or just copy our MapKit framework in the web.

We double click the SimpleMapViewController.h and start to write codes. Program Appendix B.6a shows the code in SimpleMapViewController.h.

Program Appendix B.6a SimpleMapViewController.h

```
#import <UIKit/UIKit.h>
#import <MapKit/MapKit.h>

@interface SimpleMapViewController : UIViewController {
     IBOutlet MKMapView *mapView;
     IBOutlet UIButton *hybridBtn;
     IBOutlet UIButton *standardBtn;
}

@property (nonatomic,retain) IBOutlet MKMapView *mapView;
@property (nonatomic,retain) IBOutlet UIButton *standardBtn;
@property (nonatomic,retain) IBOutlet UIButton *hybridBtn;

-(IBAction)locate;
-(IBAction)hkLocate;
-(IBAction)btnPressed:(id)sender;

@end
```

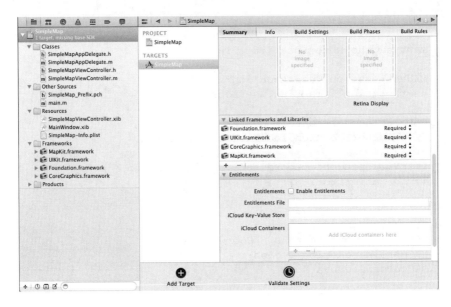

Figure B.23   Add MapKit framework in Xcode.

B.5.1   Map Function from MapKit Frameworks

#import <MapKit/MapKit.h> imports a header file of MapKit framework which includes a lot of classes for manipulating the map function.

IBOutlet MKMapView *mapView; creates an instance of MKMapView object. An MKMapView object provides an embeddable map interface, similar to the one provided by the Google Maps application. This time we can create two UIButtons for control and three methods of IBActions return type for the event-driven by buttons. Usually, one method is enough to control all button events. However, if there are too many buttons, the method will contain a lot of lines. Here we demonstrate two different techniques to handle button events.

We decide to use -(IBAction)locate; method to handle the event driven by 'Locate' button and use -(IBAction)hkLocate; method to handle the event driven by 'HK Location' button.

-(IBAction)btnPressed:(id)sender; method receives a id type value called sender parameter. Usually, id in here is to determine which button triggers the event. At this time, we decide to use it to handle events driven by both 'Standard' and 'Hybrid' buttons.

We finish the explanation in the header file. Let's look at the SimpleMapViewController.c.

Program Appendix B.6b SimpleMapViewController.m

```
#import ''SimpleMapViewController.h"

@implementation SimpleMapViewController
@synthesize mapView, hybridBtn,standardBtn;
```

```
-(IBAction)locate{

    mapView.showsUserLocation = TRUE;

    MKUserLocation *userLocation;
    userLocation = mapView.userLocation;

    CLLocationCoordinate2D centerlocation =
        userLocation.location.coordinate;

    [mapView setCenterCoordinate:centerlocation
        animated:YES];

}
-(IBAction)hkLocate{
    CLLocationCoordinate2D hkCoordinate;
    MKCoordinateRegion region;
    MKCoordinateSpan span;

    hkCoordinate.latitude = 22.2502;
    hkCoordinate.longitude = 114.2082;

    span.latitudeDelta=0.2;
    span.longitudeDelta=0.2;

    region.span=span;
    region.center=hkCoordinate;

    [mapView setRegion:region] ;
}
-(IBAction)btnPressed:(id)sender{
    if (sender==standardBtn) {
        mapView.mapType = MKMapTypeStandard;
        standardBtn.hidden = TRUE;
        hybridBtn.hidden = FALSE;
    }else {
        mapView.mapType = MKMapTypeHybrid;
        standardBtn.hidden = FALSE;
        hybridBtn.hidden = TRUE;
    }
}

- (void)didReceiveMemoryWarning {
    [super didReceiveMemoryWarning];
}

- (void)viewDidUnload {
}
```

```
- (void)dealloc {
    [mapView dealloc];
    [super dealloc];
}

@end
```

The main program seems to be little bit long, but actually, it only consists of three major parts, handling of 'Locate,' 'HK Location' and 'Hybrid & Standard' button events.

B.5.2   Locate Yourself and Shift Center View in the Map

Let's first state what we expect for when we click the 'Locate' button. We assume the device should locate where you are in the map and center the map according to your location. But it is in an iPhone simulator. It only gets the user location from the current settings of Mac machine.

```
-(IBAction)locate{

    mapView.showsUserLocation = TRUE;

    MKUserLocation *userLocation;
    userLocation = mapView.userLocation;

    CLLocationCoordinate2D centerlocation =
            userLocation.location.coordinate;

    [mapView setCenterCoordinate:centerlocation
            animated:YES];
}
```

mapView.showsUserLocation=TRUE is to assign the TRUE value for displaying the user location. After assigning to be true, the instance of an MKMapView object will locate the user location by obtaining iPhone OS system setting, which includes the current location. The MKUserLocation class defines a specific type of annotation that identifies the user's current location. You do not create instances of this class directly. Instead, you retrieve an existing MKUserLocation object from the userLocation property of the map view displayed in your application. Therefore, we set the userLocation by mapView.userLocation.

mapView.userLocation is the read-only annotation object representing the user's current location. CLLocationCoordinate2D is a structure that contains geographical coordinate. We declare a private variable, centerlocation to store the userLocation.location.coordinate value.

[mapView setCenterCoordinate:centerlocation animated:YES]; method shift the map view to current user location defined by centerlocation with animation.

B.5.3   Translate and Zoom by MKCoordinateRegion Class

Secondly, when we click to 'HK Location' button, the program should show where Hong Kong is in the map. Because the area of Hong Kong is small, if we see it macroscopically, we may not be able to identify it. So, we need to dive into some levels of details in order to locate it. Let's look at -(IBAction)hkLocate method.

```
-(IBAction)hkLocate{
    CLLocationCoordinate2D hkCoordinate;
    MKCoordinateRegion region;
    MKCoordinateSpan span;

    hkCoordinate.latitude = 22.2502;
    hkCoordinate.longitude = 114.2082;

    span.latitudeDelta=0.2;
    span.longitudeDelta=0.2;

    region.span=span;
    region.center=hkCoordinate;

    [mapView setRegion:region] ;
}
```

Inside -(IBAction)hkLocate method, we set the latitude and longitude by the following codes:

```
    hkCoordinate.latitude = 22.2502;
    hkCoordinate.longitude = 114.2082;
```

MKCoordinateRegion is a structure that defines which portion of the map to display. MKCoordinateSpan is a structure that defines the area spanned by a map region.

span.latitudeDelta=0.2; sets the amount of north-to-south distance to be 0.2 to display on the map. Similarly, span.longitudeDelta=0.2; sets the amount of east-to-west distance to be 0.2 to display for the map region. These two parameters are used to adjust the level of details.

region.center=hkCoordinate; sets the center region to be Hong Kong. [mapView setRegion:region] ; sets the display region by passing the argument region.

B.5.4   Switch from Satellite Map to Standard Street Map

Finally, we would like to display the map with satellites' pictures or a standard plain view. Two buttons, 'Hybrid' and 'Standard' serve as opposite functions, just like a switch. By default, the map is displayed the standard way, so we should hide the 'Standard' button. When we click 'Hybrid' button, the program shows up the map with the satellites' pictures. Then, the 'Standard button' should be swapped into the front view and the 'Hybrid' button should be hidden, vice versa. We use the following codes to achieve the goal.

```
-(IBAction)btnPressed:(id)sender{
    if (sender==standardBtn) {
        mapView.mapType = MKMapTypeStandard;
        standardBtn.hidden = TRUE;
        hybridBtn.hidden = FALSE;
    }else {
        mapView.mapType = MKMapTypeHybrid;
```

```
        standardBtn.hidden = FALSE;
        hybridBtn.hidden = TRUE;
    }
}
```

-(IBAction)btnPressed:(id)sender method helps to handle the button driven and pass the id (called sender) of which button is pressed into this method. We use if-else case to determine which button is pressed and which button is required to hide.

mapView.mapType = MKMapTypeHybrid; is to display the map with satellites' pictures and standard view. Changing the value in this property may cause the receiver to begin loading new map content. For example, changing from MKMapTypeStandard to MKMap-TypeSatellite might cause it to begin loading the satellite imagery needed for the map. If new data is needed, however, it is loaded asynchronously and appropriate messages are sent to the receiver's delegate indicating the status of the operation.

After we define the method, we need to link together the variables or parameters in the code and user interface in the **Interface Builder**. Once again, we click SimpleMapViewer-Controller.xib to launch Interface Builder. We click the File's Owner in SimpleMapView-erController.xib window as shown in Figure B.24(a).

Then we look at the Simple Map View Controller Connections window. We start to drag the relationship to the co-related buttons and views. When we drag into the button, we choose 'Touch Up Inside.' Noted that btnPressed will be dragged to both 'Hybrid' and 'Standard' button. Finally, you should have these similar connections as shown in Figure B.25.

Now, everything is done. Click 'Build and Run' button, if there is no error, iPhone simulater with the Map application will appear. Figure B.26(a) shows the output result of the Map application. Figure B.26(b) shows the map with satellites' pictures when the 'Hybrid' button is pressed. Figure B.26(c) shows the map with satellites' pictures and shifts the center to the USA when the 'Localize' button is pressed. Figure B.26(d) shows the Hong Kong map with satellite pictures when the 'HK Location' button is pressed. Figure B.26(e) shows the Hong Kong map with the standard view when the 'Standard' button is pressed. Figure B.26(f) shows the zooming in and out function when you press Alt key to drag. If you are able to do

(a) Choose file's owner.

(b) Simple map view controller connections window.

Figure B.24   View controller.

Figure B.25   Dragging from btnPressed to 'Hybrid' buttons.

what has been shown in Figures B.26(a)–B.26(f), you will have successfully implemented a simple map application in iPhone. We could add more functions based on this application. Let's add a slider to adjust the zoom level of the map.

B.5.5   UISlider Item Handles Zoom Events

The last map application is difficult to zoom out by the mouse cursor. We can make use of a powerful UI component, UISlider, to adjust the zoom level of the map. We base this on the previous program and open the **Interface Builder** by double clicking SimpleMapViewerController.xib in Xcode. We add UISlider in the **Library** window and drag it to **View** window. You can sketch the length of UISlider as shown in Figure B.27.

We need to change the value of UISlider from 0.5 to 0.01 as shown in Figure B.27 because, at the very beginning, the map has a low level of detail. We click the UISlider and change the value in the **Slide Attributes** window.

We go back to Xcode and add codes in SimpleMapViewerController.h.

Program Appendix B.7a SimpleMapViewController.h

```
#import <UIKit/UIKit.h>
#import <MapKit/MapKit.h>

@interface SimpleMapViewController : UIViewController {
    IBOutlet MKMapView *mapView;
    IBOutlet UIButton *hybridBtn;
    IBOutlet UIButton *standardBtn;
    IBOutlet UISlider *mapSlider;
}
```

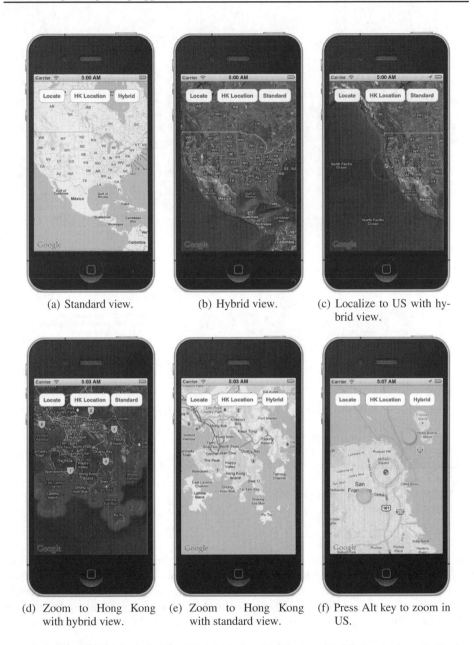

(a) Standard view.   (b) Hybrid view.   (c) Localize to US with hybrid view.

(d) Zoom to Hong Kong   (e) Zoom to Hong Kong   (f) Press Alt key to zoom in
    with hybrid view.   with standard view.   US.

Figure B.26 Different types of view. (A full color version of this figure appears in the color plate section.)

```
@property (nonatomic,retain) IBOutlet MKMapView *mapView;
@property (nonatomic,retain) IBOutlet UIButton *standardBtn;
@property (nonatomic,retain) IBOutlet UIButton *hybridBtn;
@property (nonatomic,retain) IBOutlet UISlider *mapSlider;
```

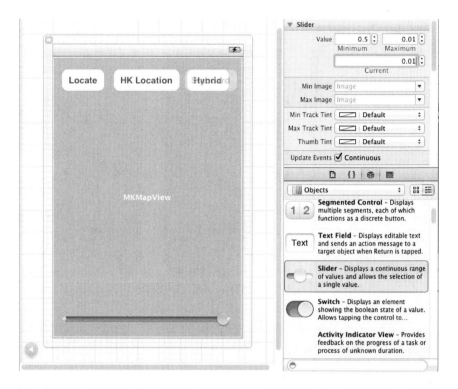

Figure B.27   UISlider item in View Window and initial value of UISlider as 0.01.

```
-(IBAction)locate;
-(IBAction)hkLocate;
-(IBAction)btnPressed:(id)sender;
-(IBAction)zoom;
```

```
@end
```

IBOutlet UISlider *mapSlider; is to add an instance of the UISlider object. A UISlider object is a visual control used to select a single value from a continuous range of values. Sliders are always displayed as horizontal bars. An indicator, or thumb, notes the current value of the slider and can be moved by the user to change the setting. We also add a method -(IBAction)zoom; to control the zooming effect of the map by the UISlider instance.

Program Appendix B.7b -(IBAction)zoom; method in SimpleMapViewController.m

```
-(IBAction)zoom{
    MKCoordinateRegion region;
    MKCoordinateSpan span;

    span.latitudeDelta=125*(1-mapSlider.value)+0.01;
    span.longitudeDelta=0.001;
```

```
region.span=span;
region.center=mapView.centerCoordinate;

[mapView setRegion:region ] ;
```

}

-(IBAction)zoom; method implements the method to zoom in and out of the map by adjusting the UISlider instance. mapSlider.value gives the adjusting value of the UISlider instance from 0 (leftmost) to 1 (rightmost). Let's see how we can adjust the zoom level by following codes:

```
span.latitudeDelta=125*(1-mapSlider.value)+0.01;
span.longitudeDelta=0.001;
```

As mentioned, span.latitudeDelta is the amount of north-to-south distance and span.longitudeDelta is the amount of east-to-west distance to display for the map region. The larger the value of span.latitudeDelta is the less detail of the map there is (zooming out). If we adjust both span.longitudeDelta and span.latitudeDelta by mapSlider.value, the map will shift the original focus. Therefore, we only adjust span.latitudeDelta. We need to keep span.latitudeDelta and span.longitudeDelta to be more than 1, such that the program will run without division by zero error.

region.center=mapView.centerCoordinate; sets the focus to current center of the screen. After we define this method, we go back to **Interface Builder** and link them the **View** and **Controller** codes together. We double click SimpleMapViewController.xib and select File's Owner. We drag mapSlider and zoom both to UISlider as shown in Figure B.28(a).

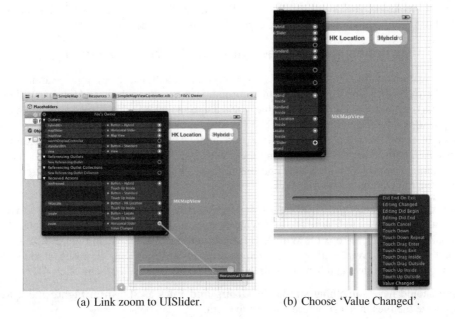

(a) Link zoom to UISlider. (b) Choose 'Value Changed'.

Figure B.28 UISlider interface.

(a) Zoom to San Francisco.   (b) Zoom to a Hong Kong   (c) Zoom to China.
                             City.

Figure B.29   Zoom to different area. (A full color version of this figure appears in the color
plate section.)

After we drag it, a pull-down menu appears and we select 'Value Changed.' 'Value Changed'
is set to capture the varying value. Save and exit the **Interface Builder**.

Finally, we click 'Build and Run.' Figure B.29(a)–B.29(c) show results of adjusting the
UISlider. Congratulations! You have learnt how to use UISlider.

B.5.6   Switches Web Browser and Simple Map Application

A navigation application usually consists of multiple functions, for example searching information on the web, displaying traffic information and locating the user. However, iPhone
can run only one application at once due to the resource limitation. If we need to handle two
applications, we usually use UITableViewController class to switch among them. Let's do
a simple exercise. We combine previous two applications, web browser and simple map,
into one application and switch among them with UITableViewController class. As usual,
we start a new project with Xcode and this time we choose **Single View Application**. We
save the file with the name MapWeb.

We open Xcode and right-click the folder of **Classes** in **Groups and Files**. Choose to add
a new file as shown in Figure B.30. A window appears, please choose 'UIViewController
subclass' and click the box of 'With XIB for user interface' as shown in Figure B.31. Click
Next button and save the file with the name, SimpleMapViewController. Repeat the same
step, we add and save another file with the name, WebBrowserViewController.

You should have something similar in **Group and Files** as shown in Figure B.32. Re-
organize the files in list structure and group them.

We click to RootViewController.h and start to code on it.

Figure B.30   Choose 'New File.'

Figure B.31   Add new class.

Figure B.32   Group and files.

Program Appendix B.8a RootViewController.h

```
#import ''SimpleMapViewController.h"
#import ''WebBrowserViewController.h"

@interface RootViewController : UITableViewController {
    NSMutableArray *tableList;
}

@end
```

The UITableViewController class creates a controller object that manages a table view. In this program, we use NSMutableArray instance to store multiple views and controllers. The NSMutableArray class declares the programmatic interface to objects that manage a modifiable array of objects. Our main goal is to provide a root table list to switch between two applications, SimpleMap and WebBrowser. We import these two application views of header files in RootViewController header file which helps to switch among three views, RootView, SimpleMapView and WebBrowserView.

Program Appendix B.8b RootViewController.c

```
#import ''RootViewController.h"

@implementation RootViewController

- (void)viewDidLoad {
    [super viewDidLoad];

    tableList = [[NSMutableArray alloc] init];
    [tableList addObject:@"Simple Map Application"];
    [tableList addObject:@"Web Browser"];
    [self setTitle:@"Navigation Item List"];
}

- (void)didReceiveMemoryWarning {
    [super didReceiveMemoryWarning];
}

#pragma mark Table view methods

- (NSInteger)numberOfSectionsInTableView:(UITableView *)
    tableView {
    return 1;
}

- (NSInteger)tableView:(UITableView *)tableView
    numberOfRowsInSection:
(NSInteger)section {
    return [tableList count];
}
```

```objc
- (UITableViewCell *)tableView:(UITableView *)tableView
    cellForRowAtIndexPath:
    (NSIndexPath *)indexPath {

    static NSString *CellIdentifier = @"Cell";

    UITableViewCell *cell = [tableView
        dequeueReusableCellWithIdentifier:CellIdentifier];
    if (cell == nil) {
        cell = [[UITableViewCell alloc]
            initWithStyle:  UITableViewCellStyleDefault
            reuseIdentifier:CellIdentifier];
    }

     cell.text=[[tableList objectAtIndex:indexPath.row] retain];
     [cell setAccessoryType:
    UITableViewCellAccessoryDisclosureIndicator];

    return cell;
}

- (void)tableView:(UITableView *)tableView didSelectRowAtIndexPath:
(NSIndexPath *)indexPath {

    if ([[tableList objectAtIndex:indexPath.row]
        isEqual:@"Simple Map Application"])
    {
        SimpleMapViewController *smvc=[[SimpleMapViewController
                alloc]
        initWithNibName:@"SimpleMapViewController" bundle:nil];
          [smvc setTitle:@"Map Application"];
          [self.navigationController pushViewController:smvc
        animated:YES];
    }

    if ([[tableList objectAtIndex:indexPath.row]
        isEqual:@"Web Browser"])
    {
        WebBrowserViewController *wbvc=[
          [WebBrowserViewController alloc]
          initWithNibName:@"WebBrowserViewController"
          bundle:nil];
         [wbvc setTitle:@"Web Browser"];
         [self.navigationController
        pushViewController:wbvc
        animated:YES];
    }
}
```

```
- (void)dealloc {
    [super dealloc];


}
@end
```

RootViewController.m is quite long in context. Let's break it down and discuss some important aspects. When the root view is loaded, the menu should be ready with the name of each view loaded. The following codes do the above task:

```
- (void)viewDidLoad {
    [super viewDidLoad];

    tableList = [[NSMutableArray alloc] init];
    [tableList addObject:@"Simple Map Application"];
    [tableList addObject:@"Web Browser"];
    [self setTitle:@"Navigation Item List"];
}
```

[tableList addObject:@'Simple Map Application']; helps to add a NSMutableArray object called 'Simple Map Application.' We add two NSMutableArray objects and set their names.

```
- (NSInteger)tableView:(UITableView *)
    tableView numberOfRowsInSection:
    (NSInteger)section {
    return [tableList count];
}
```

return [tableList count]; returns the number of a NSMutableArray object into the (NSInteger)tableView: method.

cell.text=[[tableList objectAtIndex:indexPath.row] retain]; to set the name of each UITableViewCell instance in the list.

[cell setAccessoryType:UITableViewCellAccessoryDisclosureIndicator]; provides a small icon of forward click. You could replace the small icon by changing the following code: [cell setAccessoryType:UITableViewCellAccessoryDetailDisclosureButton];

The following code lines are to react when an NSMutableArray object is selected.

```
- (void)tableView:(UITableView *)tableView
    didSelectRowAtIndexPath: (NSIndexPath *)
    indexPath {

    if ([[tableList objectAtIndex:indexPath.row]
        isEqual:@"Simple Map Application"])
    {
        SimpleMapViewController *smvc=[
            [SimpleMapViewController alloc]
            initWithNibName:@"SimpleMapViewController"
            bundle:nil];
        [smvc setTitle:@"Map Application"];
        [self.navigationController
```

(a) Application list.　　(b) Map Application under　(c) Web Browser under the
　　　　　　　　　　　　　　the list.　　　　　　　　　list.

Figure B.33　Application list.

```
    pushViewController:smvc animated:YES];
}

if ([[tableList objectAtIndex:indexPath.row]
    isEqual:@"Web Browser"])
{
    WebBrowserViewController *wbvc=[
    [WebBrowserViewController alloc]
    initWithNibName:@"WebBrowserViewController"
    bundle:nil];
        [wbvc setTitle:@"Web Browser"];
        [self.navigationController
    pushViewController:wbvc animated:YES];
}
}
```

if ([[tableList objectAtIndex:indexPath.row] is Equal:@'Simple Map Application']) If-case
is used to determine which an NSMutableArray object is selected by a NSString.

Instead of using Interface Builder, we use the line to choose the view of controller.

```
SimpleMapViewController *smvc=[
[SimpleMapViewController alloc]
initWithNibName:@"SimpleMapViewController"
bundle:nil];
```

[self.navigationController pushViewController:smvc animated:YES]; helps to create push-
ing animation when the application is switched.

After we finish editing RootViewController.h and RootViewController.m files, we set and code to SimpleMap and WebBrowser again. If everything goes smoothly, you will have following results as shown in Figure B.33.

Chapter Summary

In this appendix chapter, we've introduced the user interface of Xcode and how to run and compile in a console environment. Then, we've gone through several new features of Objective-C language. They are @interface, @implementation, @property, @synthesize, @dynamic, infix notation, dot notation and category. The @interface section is usually in a header file that initially defines the class. The @implementation section is to define methods concretely. @property in the @interface section is to declare to have getter and setter of a variable. @synthesize in the @implementation section is to define the getter and setter automatically. @dynamic in the @implementation section allows us to define the getter and setter manually. Infix notation is used in Objective-C language and follows [ClassOrInstance method: argument]. Dot notation is used to access object attributes in Objective-C. Category helps to add more functions into the existing class without sub-classing.

Also, we started to talk about iPhone programming by a classic HelloWorld program. We have introduced the use of Interface Builder, Library Tool, etc. The MVC programming development model has been covered. Later on, we programmed web browser and simple map application. In the meantime, you have learnt how to use some UIComponents such as UIButton, UIWindow, UIView, UIWebView and UISlider, etc. We initially learned some basic functions of MapKit framework and applied viewing, locating and zooming function of map.

Index

accelerometer, 79
acoustic-based positioning, 186
active acoustic positioning, 186, 187
active RFID Tag, 195
adaptive beamformer, 188
adaptive video streaming, 228–229
age of data offset, 140
almanac, 126, 138, 141, 167
angle of arrival, 38–40, 195
anti-spoofing, 124
anti-spoofing indicator, 139
Apple80211 framework, 74
Apple's HTTP Live Streaming Protocol, 227,
    229
artificial intelligence, 204
assisted global positioning system, 2, 167
atmospheric errors, 129

back-propagation neural network, 52
basic service set identification, 74
Bayes' algorithm, 58
beamforming, 187
bessel function, 65
binary offset carrier, 143
broadcast video streaming, 225

camera-based positioning, 188
cellular positioning System, 2
channel assignment, 23
channel interference, 20
Chi-square distribution, 21
circular error of probable, 58, 63
clock bias, 146
clutter density factor, 15
coarse acquisition code, 124, 138
control segment, 126
coordinated universal time, 129, 141
core location framework, 168
Cramèr-Rao lower bound, 58, 59

cross-correlation technique, 43
cumulative distribution function, 58, 59

differential global positioning system, 157
dilution of precision, 131
discrete Fourier transformation, 102
Doppler frequency, 154

Eigen-space, 191
energy consumption model, 186
energy loss algorithm, 186
environmental factor, 24, 30
ephemerides, 127
ephemeris, 126, 138
ephemeris errors, 128
ephermeris, 167
Euler's identity, 102

fast Fourier transform, 101
feature-based recognition, 191
feature selection, 207
Federal Aviation Administration, 159
Fisher information matrix, 61
Fisher-space approach, 191
Fourier coefficients, 101
Fourier descriptor, 101
fuzzy color map, 108
fuzzy logic, 205
fuzzy membership function, 108, 206
fuzzy membership graph, 206

general packet radio service, 227
geodetic coordinate system, 133
geometric dilution of precision, 58, 65, 131
global navigation satellite system, 165
global positioning system, 1, 123
GPS receiver, 127
great circle distance, 135
ground antenna, 127

Introduction to Wireless Localization: With iPhone SDK Examples, First Edition. Eddie C.L. Chan and
George Baciu.
© 2012 John Wiley & Sons Singapore Pte. Ltd. Published 2012 by John Wiley & Sons Singapore Pte. Ltd.